The Audi TT

A Journey Through Design and Innovation

Todd Bandel

Copyright © 2024 Todd Bandel

All rights reserved.

ISBN: 9798304729185

DEDICATION

I dedicate this book to all my past Audi automotive mentors and colleagues. Your guidance, support, and shared wisdom have been invaluable in shaping my journey. Each of you played a significant role in my professional development, imparting knowledge and fostering a passion for excellence in the automotive field.

CONTENTS

ACKNOWLEDGEMENT   I

Chapter One
The Birth of an Icon: Concepting the Audi TT   1

Chapter Two
From Sketch to Reality: The Design Evolution   13

Chapter Three
Engineering Excellence: The TT's Technical Foundations   27

Chapter Four
The First Generation: Revolutionizing Sports Car Design   43

Chapter Five
Performance and Handling: The TT's Driving Dynamics   57

Chapter Six
The TT in Motorsports: Racing Heritage and Achievements   73

Chapter Seven
The Second Generation: Refining a Classic   89

Chapter Eight
Pushing Boundaries: TT RS and High-Performance Variants   105

Chapter Nine
The Third Generation: Embracing Modern Technology   119

Chapter Ten
Inside the TT: Interior Innovations and Comfort   131

Chapter Eleven
The TT's Cultural Impact   149

Chapter Twelve
Looking Ahead: The Future of the Audi TT in the Electric Era   159

ACKNOWLEDGMENTS

I want to express my deepest gratitude to my father for introducing me to the exhilarating world of automotive racing. Your passion for cars and dedication to the sport have inspired me. From the first time you brought me to a race track, I was captivated by the power and precision of the machines and the skill required to master them. Your guidance and support have fueled my interest and enthusiasm, making every moment in this thrilling world more meaningful. Thank you for sharing this incredible journey with me and being such a pivotal influence in my life.

The Audi TT: A Journey Through Design and Innovation

Chapter 1: The Birth of an Icon: Concepting the Audi TT

Section 1.1: The Spark of Inspiration

In the annals of automotive history, few origin stories are as captivating as that of the Audi TT. The spark that ignited this design revolution occurred during a fateful meeting at Volkswagen Group in 1994. As the clock ticked late into the night, designers Freeman Thomas and J Mays found themselves in the midst of an intense brainstorming session. With nothing but a napkin and a shared vision, they began to sketch the first contours of what would become one of the most iconic sports cars of the modern era.

The influence of Bauhaus design principles was crucial in shaping the TT's concept. The clean lines and geometric shapes that would come to define the TT's aesthetic can be traced back to the Bauhaus philosophy of "form follows function." This approach, which emphasizes simplicity and functionality, provided the perfect framework for creating a sports car that was both visually striking and purposeful in its design.

The Audi TT: A Journey Through Design and Innovation

The name "TT" itself carries significant weight in the automotive world. Far from being a random combination of letters, it pays homage to two essential elements of Audi's rich history. The "TT" moniker is a nod to the NSU TT. This small yet mighty car made a significant impact in the 1960s, as well as at the Isle of Man TT race, a grueling motorcycle event long associated with speed and precision – qualities the designers hoped to embody in their new creation.

As Thomas and Mays continued to refine their vision, it became clear they aimed to create something truly unique in the sports car market. Their goal was to design a vehicle that would turn heads and quicken pulses without the astronomical price tag typically associated with high-end sports cars. This democratization of design excellence was a bold move, one that would ultimately help to redefine Audi's place in the automotive landscape.

Initial reactions to the TT concept within Audi and the broader Volkswagen Group were mixed. Some executives were skeptical, viewing the design as a too-radical departure from Audi's traditional offerings. However, it was Ferdinand Piëch, then CEO of Volkswagen Group, who recognized the potential of this revolutionary design. With his keen eye for innovation and his willingness to take calculated risks, Piëch gave the project the green light, setting in motion a series of events that would change the face of automotive design.

As word of the project spread through the halls of Audi and Volkswagen, it began to attract a team of visionary designers and engineers who were eager to be part of something groundbreaking. The excitement was palpable as the concept began to take shape, with each team member bringing their unique skills and perspectives to the table.

The initial sketch that started it all quickly evolved into more detailed drawings and eventually into clay models. With each iteration, the TT's shape became more refined, its character more defined. The team worked tirelessly to preserve the purity of the original concept, fighting against the compromises that often water down bold designs as they move towards production.

As news of the project began to leak out to the automotive press, anticipation started to build. Rumors swirled about a new Audi sports car that would redefine the genre. The team felt the weight of these expectations, but rather than being daunted, they were inspired to push even harder, to create something truly extraordinary.

The spark of inspiration that was ignited in that late-night meeting had grown into a roaring fire of creativity and innovation. The Audi TT was no longer just a concept; it was rapidly becoming a reality that would soon set the automotive world ablaze. Little did anyone know at the time, but this was the beginning of a journey that would not only transform Audi but would leave an indelible mark on the entire automotive industry.

Section 1.2: From Concept to Reality

The journey from initial sketch to production-ready vehicle was a testament to the dedication and vision of the Audi design team. At the helm of this ambitious project was a group of talented individuals, each bringing their unique expertise to the table. Peter Schreyer, who would later become one of the most renowned automotive designers in the industry, joined the team to refine the TT's shape. His keen eye for detail and innovative approach to design played a crucial role in maintaining the concept's bold aesthetic while adapting it for real-world production.

Unlike many concept cars that get watered down for production, the team behind the TT was determined to preserve its bold design elements. This commitment to the original vision was not without its challenges. The rounded roof, while visually striking, posed unique challenges for structural integrity and manufacturing. Engineers worked tirelessly to develop innovative solutions that would make this signature feature a reality without compromising safety or functionality.

The Audi TT: A Journey Through Design and Innovation

The TT's development coincided with a revolution in automotive design technology. It was one of the first cars to be designed almost entirely using computer-aided design (CAD) tools. This cutting-edge approach enabled greater precision and faster iterations, allowing the team to experiment with various design elements and refine the car's shape with unprecedented accuracy. The use of CAD technology also facilitated closer collaboration between designers and engineers, ensuring that the TT's bold aesthetics could be translated into a fully functional vehicle.

Throughout the development process, the team faced the constant challenge of balancing form and function. While the exterior was undoubtedly innovative, equal attention was paid to the interior. The designers ensured that the cabin was not only visually appealing but also ergonomically sound for drivers. Every element, from the placement of controls to the shape of the seats, was carefully considered to create a driving experience that matched the car's striking external appearance.

The team's unwavering commitment to the TT's original concept was driven by a shared belief in the power of design to evoke emotion and challenge conventions. They recognized that the TT had the potential to be more than just another sports car; it could be a statement piece that would redefine Audi's image and influence the entire automotive industry.

As the concept inched closer to production reality, excitement grew within Audi. The TT was poised to become a flagship model that would showcase the brand's design prowess and technological capabilities. The journey from concept to reality was a testament to the power of vision, teamwork, and innovative thinking – elements that would become hallmarks of the Audi TT's enduring legacy.

Section 1.3: The Concept Car Debut

The process from initial sketch to fully realized concept car culminated in one of the most anticipated moments in automotive history: the debut of the Audi TT concept at the 1995 Frankfurt Motor

Show. The lead-up to this momentous occasion was filled with excitement and nervous energy as the design team worked tirelessly to perfect every detail of their trailblazing creation.

As the day of the unveiling approached, the Frankfurt Motor Show buzzed with anticipation. Rumors had been circulating about a revolutionary new concept from Audi, and industry insiders were eager to see what the German automaker had in store. The Audi stand was meticulously prepared, with the TT concept hidden beneath a sleek cover, ready for its grand debut.

When the moment finally arrived, a hush fell over the crowd as Audi executives took the stage. As the covers were pulled back, revealing the TT's shape for the first time, there was a moment of stunned silence. This was quickly followed by a burst of excitement as journalists and industry professionals surged forward to get a closer look at the car that would redefine sports car design.

The initial reaction was overwhelmingly positive. Auto magazines hailed the TT as a bold step forward in sports car design, with one critic calling it "a concept car come to life." The TT's clean lines, attractive silhouette, and innovative use of materials captured the imagination of both the press and the public. Many were struck by how the design seamlessly blended elements of retro styling with futuristic aesthetics, creating something entirely new and captivating.

The impact of the TT concept's debut extended far beyond the Audi stand. As news of the breathtaking design spread throughout the show, competitors scrambled to re-evaluate their own sports car designs. The TT had set a new benchmark for what a modern sports car could look like, and other manufacturers knew they would need to up their game to compete.

The most significant outcome of the Frankfurt debut was Audi executives' decision to fast-track the TT for production. The overwhelmingly positive response convinced them that they had a potential game-changer on their hands. This decision would prove pivotal for Audi, setting the stage for a new era of design leadership within the company.

The concept car's debut also marked a shift in public perception of the Audi brand. No longer seen as just a maker of reliable luxury sedans, Audi was now viewed as a cutting-edge design leader capable of producing fascinating sports cars. This shift would have far-reaching implications for the brand's future product strategy and marketing efforts.

As the Frankfurt Motor Show came to a close, the automotive world was abuzz with talk of the Audi TT concept. It had achieved what few concept cars manage to do: capture the imagination of both industry insiders and the general public, while also signaling a bold new direction for its parent company. The journey from a sketch to a show-stopping concept car was complete, but for the Audi TT, this was just the beginning.

Section 1.4: Refining the Concept for Production

The journey from concept to production is often where the true challenges of automotive design reveal themselves. For the Audi TT, this process was a delicate balance between preserving the bold vision of the concept car and adapting it to the realities of mass production and real-world use.

One of the first considerations in refining the TT for production was making subtle design tweaks for manufacturability. While the overall shape that had captivated audiences at the Frankfurt Motor Show remained intact, the design team made minor adjustments to ensure the car could be produced efficiently and meet various safety regulations. For instance, the side mirrors were slightly enlarged and reshaped to improve visibility while retaining their design. The bumpers were also subtly modified to incorporate impact-absorbing materials without compromising the car's sleek silhouette. These changes were so skillfully executed that to the casual observer, the production TT appeared virtually identical to the concept.

The Audi TT: A Journey Through Design and Innovation

Choosing the mechanical underpinnings for the TT was another crucial decision in the refinement process. Audi opted to base the TT on the Volkswagen Group A platform, which was also used in the Audi A3. This decision was driven by a desire to balance performance with cost-effectiveness. By utilizing an existing platform, Audi could keep development costs in check while still delivering the driving dynamics expected of a sports car. The choice also allowed greater flexibility in drivetrain options, enabling Audi to offer both front-wheel-drive and quattro all-wheel-drive variants of the TT.

The interior design of the TT received just as much attention as the exterior during the refinement process. The team was committed to carrying the minimalist exterior theme into the cabin. This approach resulted in several elements that would become instant classics. Perhaps most emblematic were the aluminum-ringed air vents, which echoed the circular motif found throughout the car's design. The instrument panel was designed with a driver-centric layout, emphasizing the TT's sporting intentions. Every element, from the shape of the steering wheel to the placement of controls, was carefully considered to create a cohesive, functional interior space that is as visually striking as the exterior.

Color and material selection played a crucial role in complementing the TT's design. The team curated a palette of colors and materials that emphasized the car's futuristic look while also offering options to suit a variety of tastes. One of the most memorable choices was introducing the "Baseball Leather" interior option. This unique treatment featured leather seats with stitching that mimicked the appearance of a baseball glove, adding a touch of whimsy and craftsmanship to the TT's interior. Exterior color choices were equally considered, with a range of hues selected to accentuate the car's lines and curves.

Before the TT could enter production, it underwent a rigorous testing process to ensure it could deliver on its promises of performance and quality. Prototypes were subjected to extensive wind tunnel testing to verify that the car's shape performed well aerodynamically. This testing not only helped optimize the car's efficiency and high-speed stability but also ensured that the unique design elements, such as the rounded roof and truncated rear, didn't create unexpected air flow issues. Road tests were conducted in various climates and conditions to fine-tune the suspension and drivetrain, ensuring the TT delivered a driving experience worthy of its sporty appearance.

Throughout the refinement process, the team remained committed to preserving the essence of the original concept. This dedication to the initial vision is what sets the TT apart from many other concept cars that become diluted on their way to production. The result was a production car that not only lived up to the excitement generated by the concept but, in many ways, exceeded it, proving that bold, avant-garde design could be successfully translated into a functional, mass-produced vehicle. The refined Audi TT stood as a testament to the power of staying true to a visionary concept while skillfully adapting it to the demands of the real world.

Section 1.5: The Impact of the TT Concept

The Audi TT concept car left an indelible mark on the automotive industry, influencing not just Audi's future but the entire landscape of sports car design. Its impact was far-reaching and multifaceted, touching on various aspects of the automotive world.

One of the most significant effects of the TT concept was its role in shifting Audi's brand image. Before the TT, Audi was primarily known for producing reliable, if somewhat conservative, sedans. The unveiling of the TT concept car at the 1995 Frankfurt Motor Show marked a turning point. Its bold, avant-garde design signaled to the world that Audi could create cutting-edge, emotionally stirring vehicles. This concept car single-handedly transformed Audi's image from a maker of dependable cars to a leader in automotive design and innovation.

The influence of the TT concept extended well beyond Audi's brand perception. Its design elements began to permeate the broader automotive design world. The TT's clean lines, wheel arches, and minimalist aesthetic started appearing in other manufacturers' designs within a few years of its debut. Elements such as the TT's roofline and the way its body seemed to be milled from a single piece of metal became inspirations for designers across the industry. The TT concept didn't just create a new car; it created a new design language that would influence sports cars and even some sedans and SUVs for years to come.

One of the most lasting impacts of the TT concept was raising expectations for fidelity from concept to production. In an era where concept cars often bore little resemblance to their eventual production versions, the TT broke the mold. Audi's commitment to bringing the TT to market with minimal changes from the concept version set a new industry standard. This successful transition from concept to production car encouraged other manufacturers to be bolder with their concept designs and to strive harder to maintain the integrity of those designs in production models. The TT proved that a daring concept could indeed become a viable production car without losing its essence.

The TT concept also played a crucial role in attracting a new demographic to the Audi brand. Its avant-garde design and promise of accessible performance appealed to younger buyers and design enthusiasts who might not have previously considered an Audi. This new audience included creative professionals, tech entrepreneurs, and urban trendsetters drawn to TT's combination of style and substance. By appealing to this demographic, Audi rejuvenated its customer base and positioned itself as a brand for forward-thinking individuals.

Lastly, the TT concept set the stage for future Audi sports cars. Its success paved the way for more daring projects within Audi, culminating in vehicles like the R8 supercar. The positive reception of the TT concept gave Audi the confidence to push boundaries in its sports car offerings, leading to a new era of performance vehicles for the brand. The TT concept proved that Audi could compete in the sports car market, not just on performance but also on vehicle design.

In essence, the impact of the Audi TT concept car was amazing. It transformed Audi's image, influenced industry-wide design trends, set new standards for concept cars, attracted a fresh audience to the brand, and ushered in a new era of Audi sports cars. The TT concept wasn't just a preview of a new model; it was a glimpse into the future of automotive design and brand strategy. Its influence continues to resonate in the automotive world, serving as a testament to the power of visionary design and bold decision-making in shaping the future of an entire industry.

Section 1.6: The TT's Cultural Impact

The Audi TT's influence extended far beyond the automotive world, shaping popular culture and design trends across industries. From the moment it was unveiled as a concept, the TT captured the imagination of not just car enthusiasts but also artists, designers, and the general public.

In the realm of product design, the TT's clean lines and minimalist aesthetic inspired a new wave of consumer goods. Home appliances, electronics, and even furniture began to adopt a similar design language, emphasizing simplicity and geometric forms. The TT's influence could be seen in everything from sleek coffee makers to streamlined desk lamps, as designers sought to capture the same sense of modernity and timelessness that the TT embodied.

The fashion world also took notice of the TT's striking silhouette. High-end fashion designers drew inspiration from the car's curves and proportions, incorporating similar elements into their clothing and accessory designs. The TT's shape even influenced eyewear, with several popular sunglasses brands releasing models that echoed the car's bold, wraparound look.

In architecture, the TT's design principles of simplicity and clarity found their way into modern building aesthetics. Architects began to emphasize clean lines and geometric shapes in their structures, mirroring the TT's harmonious blend of form and function. The car's influence could be seen in everything from sleek urban lofts to cutting-edge corporate headquarters.

The TT also made its mark in popular media. It quickly became a symbol of sophistication and modernity in films and television shows, often appearing as the vehicle of choice for stylish, forward-thinking characters. The car's appearances in movies helped cement its status as a cultural icon, transcending its role as mere transportation to become a statement piece.

In the art world, the TT inspired numerous works across various media. Painters and sculptors were drawn to its unique form, creating pieces that celebrated its shape. Photography exhibitions dedicated to the TT's design sprouted up in galleries around the world, showcasing the car's ability to captivate even when stationary.

The digital realm wasn't immune to the TT's charm either. As computer-generated imagery (CGI) became more prevalent in advertising and entertainment, the TT's easily recognizable shape made it a favorite subject for 3D artists and animators. Its virtual presence in video games and digital art further solidified its place in contemporary visual culture.

Perhaps most notably, the TT became a case study in design schools and universities worldwide. Its development process, from concept to production, was analyzed and taught as an example of successful industrial design. The car's ability to maintain its conceptual purity while meeting real-world manufacturing and safety requirements became a benchmark for aspiring designers across various fields.

The TT's cultural impact also manifested in unexpected ways. Car clubs and enthusiast groups dedicated to the model sprung up around the globe, fostering communities and friendships among owners and admirers. These groups often organized events and rallies, turning the TT into a social phenomenon that brought people together through a shared appreciation for design and engineering.

As the years passed, the TT's influence on culture didn't wane. Instead, it evolved, with each new generation of the car adding to its legacy. Vintage first-generation TTs became collector's items, appreciated not just for their driving dynamics but as pieces of automotive art. The car's enduring appeal across multiple generations spoke to the timelessness of its original design concept.

In essence, the Audi TT transcended its role as a sports car to become a cultural touchstone. It represented a moment in time when automotive design took a bold leap forward, inspiring creative minds across numerous fields. The TT's cultural impact serves as a testament to the power of visionary design to shape the world around us far beyond its original intent.

Chapter 2: From Sketch to Reality: The Design Evolution

Section 2.1: The Birth of a Concept

In the early 1990s, a pivotal meeting at Volkswagen Group set the stage for what would become one of the most iconic sports car designs of the decade. It was during this gathering that the seed of the Audi TT was planted, a concept that would challenge conventional automotive design and redefine Audi's image.

At the heart of the brilliant idea were two visionary designers: Freeman Thomas and J Mays. In a moment of inspiration that has since become historic in automotive circles, Thomas sketched the TT's initial concept on a simple piece of paper. This seemingly modest act would change the trajectory of Audi's design language and leave an indelible mark on the automotive world.

The inspiration behind the TT's unique shape was a harmonious blend of modernist architecture and classic automotive design. Thomas and Mays drew influence from the Bauhaus movement, characterized by its emphasis on simple, functional forms.

This architectural influence is evident in the TT's clean lines, geometric shapes, and the perfect balance between form and function. The designers also paid homage to classic sports cars, particularly the rounded fenders and cockpit-like cabin reminiscent of the Porsche 356.

The name "TT" itself carries significant meaning, adding depth to the concept. It's a nod to the Isle of Man Tourist Trophy (TT) motorcycle race, one of the most challenging and prestigious events in motorsports. This connection wasn't just a clever marketing ploy; it embodied the spirit of performance, precision, and daring that the designers envisioned for the car.

When Thomas and Mays presented their concept to Audi executives, the initial reception was a mix of excitement and skepticism. The design was undoubtedly bold and unlike anything in Audi's current lineup. Some saw it as a risky departure from the brand's conservative image, while others recognized its potential to revolutionize Audi's design language and market position.

The concept's unique blend of retro and futuristic elements challenged the status quo. Its compact size and silhouette stood in stark contrast to the larger, more angular designs that dominated Audi's portfolio at the time. This tension between tradition and innovation would become a defining characteristic of the TT project.

Despite initial hesitation, Thomas and Mays' passion and conviction, combined with the undeniable visual impact of their design, ultimately won over Audi leadership. The green light was given to develop the concept further, setting in motion a process that would push the boundaries of automotive design and engineering.

As word of the project spread within Audi, it generated a buzz. Designers and engineers alike were energized by the opportunity to work on something truly exceptional. The TT concept became a rallying point for creativity and innovation within the company, inspiring teams to think beyond conventional constraints.

This initial phase of the TT's development demonstrates the power of a bold idea and the importance of visionary thinking in automotive design. From a simple sketch to a fully developed concept that captured the imagination of Audi's top brass, the birth of the TT marked the beginning of a new era for Audi. It set the stage for a design revolution that would not only transform the brand but also leave a lasting impact on the automotive industry as a whole.

Section 2.2: From Paper to Clay

The transition from two-dimensional sketches to a tangible, three-dimensional model is a crucial step in automotive design, and the Audi TT was no exception. Once Freeman Thomas and J Mays' initial concept sketch had captured the imagination of Audi's design team, the process of bringing the TT to life began in earnest.

Translating the essence of the TT's unique shape from paper to clay was a meticulous process that required both artistic vision and technical precision. The design team started by creating multiple detailed drawings, exploring various angles and proportions to refine the original concept. These drawings served as the foundation for the next phase: clay modeling.

Clay modeling has long been a cornerstone of automotive design, enabling designers to shape and refine their vision physically. For the TT, skilled modelers painstakingly crafted full-scale clay models, carefully translating every curve and line from the drawings into three dimensions. This process allowed designers to evaluate the car's proportions, stance, and overall aesthetic in a way that two-dimensional sketches couldn't.

As the clay model took shape, the iterative design refinement process began. Designers and engineers worked collaboratively, making countless minor adjustments to perfect the TT's form. They would step back, evaluate the model from various angles, and then dive back in to sculpt and reshape. This back-and-forth continued until every line and curve met their exacting standards.

A critical part of this refinement process involved wind tunnel testing. The TT's shape wasn't just about aesthetics; it needed to perform well aerodynamically. The design team worked closely with aerodynamicists, using wind tunnel data to inform subtle changes to the car's form. This collaboration ensured that the TT would not only look stunning but also slip through the air efficiently at high speeds.

Throughout this process, the designers faced the constant challenge of balancing form and function. The TT's unique shape was a bold departure from conventional sports car design, and preserving its character while meeting practical requirements was no small feat. Every decision, from the rake of the windshield to the curve of the roofline, had to serve both aesthetic and functional purposes.

The clay modeling phase also allowed the design team to experiment with different details that would define the TT's character. Elements like the wheel arches, the subtle curve of the rear, and the clean, unbroken lines of the side profile were all refined during this stage. These details, while seemingly minor, define the TT's unmistakable silhouette.

As the clay model neared its final form, it became clear that the TT was shaping up to be something truly special. The transition from paper to clay had not only preserved the essence of the original sketch but had also enhanced it, adding depth, dimension, and a sense of real-world presence that exceeded even the designers' initial vision.

The clay model served as a crucial bridge between concept and reality, allowing the design team to refine the TT's form before moving on to subsequent stages of development. It was a tangible representation of Audi's bold new design direction, setting the stage for a car that would not only turn heads on the street but also redefine expectations for a modern sports car.

The Audi TT: A Journey Through Design and Innovation

Section 2.3: The Concept Car Unveiling

The automotive world was abuzz with anticipation as the 1995 Frankfurt Motor Show approached. Audi had been tight-lipped about their new concept car, but rumors of a magnificent design had been circulating in industry circles. As the curtain finally lifted, the Audi TT concept car stood before the crowd, its bold lines and unique silhouette immediately captivating everyone present.

The TT concept car was a testament to innovative design. Its compact, rounded body was unlike anything else on the show floor. The car's seamless blend of curves and sharp angles created a harmonious yet striking visual impact. The concept's aluminum body gleamed under the spotlights, emphasizing its futuristic appearance. One of the most talked-about features was the absence of a B-pillar, which gave the car a sleek, uninterrupted side profile.

Inside, the concept car's interior was just as innovative as its exterior. The cockpit featured a minimalist design with a driver-centric layout. Circular themes dominated the cabin, from the air vents to the instrument cluster, echoing the car's exterior curves. The use of brushed aluminum accents throughout the interior created a cohesive link with the body, reinforcing the TT's unique design language.

The public and media reaction to the TT concept was nothing short of electric. Journalists scrambled to get a closer look, camera flashes lit up the Audi stand, and crowds gathered to catch a glimpse of this automotive sculpture. The TT was instantly hailed as a design triumph, with many praising its bold departure from conventional sports car aesthetics.

Compared to contemporary sports car designs of the mid-1990s, the TT stood out as a breath of fresh air. While many sports cars of the era embraced sharp, aggressive lines, the TT offered a more sophisticated, almost sensual approach to performance-car design. Its compact dimensions and clean lines were a stark contrast to the long-hooded, wide-bodied sports cars that dominated the market.

The unveiling of the TT concept car had a profound impact on Audi's brand image. Previously known primarily for its practical, well-engineered sedans and wagons, Audi suddenly found itself at the forefront of automotive design. The TT demonstrated that Audi could produce cutting-edge, emotionally appealing vehicles that could compete with the best in the world.

The concept car's reception at Frankfurt sent a clear message to Audi's management: they had a potential breakthrough on their hands. The overwhelming positive response not only validated the bold design direction but also put pressure on the company to bring a production version to market that stayed true to the concept's aesthetics.

As the motor show came to a close, the Audi TT concept car had firmly established itself as one of the stars of the event. It had set a new benchmark for automotive design and created a wave of excitement that would carry through to its eventual production. The journey from sketch to concept car was complete, but the challenge of translating this showstopper into a road-going reality was just beginning.

Section 2.4: Transitioning to Production Design

The journey from concept car to production model is often fraught with challenges, and the Audi TT was no exception. The task of maintaining the concept car's striking aesthetics while adapting the design for mass production was a delicate balancing act that required both creativity and pragmatism.

One of the primary challenges faced by Audi's design team was preserving the concept car's clean, uncluttered lines and geometric purity. The production model needed to retain the essence of the TT's unique shape while accommodating the practical requirements of a road-going vehicle. This meant making subtle adjustments to the body panels, ensuring they could be manufactured efficiently while still capturing the concept's visual appeal.

The Audi TT: A Journey Through Design and Innovation

Safety regulations played a significant role in shaping the final design of the production TT. The concept car's low-slung profile and minimalist approach to exterior elements had to be modified to meet stringent crash protection standards. This resulted in slight changes to the front and rear overhangs, along with the integration of more pronounced bumpers. Despite these necessary alterations, the design team worked tirelessly to ensure that the TT's silhouette remained intact.

The choice of materials significantly influenced the final look of the production TT. While the concept car showcased exotic materials, the production version needed to balance aesthetics with cost-effectiveness and durability. Audi opted for a combination of aluminum and steel for the body, allowing for weight savings while maintaining structural integrity. This decision not only affected the car's performance but also its visual presence, as the interplay of light on the aluminum panels became a key design feature.

Throughout the transition, preserving the TT's unique shape remained paramount. The design team focused on retaining key elements that defined the concept car's character, such as the wheel arches, the curvaceous roof line, and the bold, singular character line that ran the length of the body. These features were carefully refined to ensure they could be consistently reproduced in a mass production setting without losing their visual impact.

The headlights and taillights underwent significant development to meet safety standards and manufacturing constraints while maintaining their role as key design elements. The production model's lighting units were larger and more complex than those of the concept, but they were sculpted to complement the car's overall design language.

Interior elements also required adaptation for production. While some of the concept's more extravagant features had to be toned down, the design team worked to retain the cockpit-like feel and driver-centric layout. This involved refining the circular theme prevalent in the concept and integrating it into practical, usable controls and displays.

One of the most challenging aspects of the transition was maintaining the concept's greenhouse proportions. The production model required slightly larger side windows and a more substantial A-pillar for structural integrity and safety. However, clever design work ensured that these necessary changes did not compromise the TT's profile.

Throughout the transition to production design, Audi's team demonstrated remarkable skill in navigating the competing demands of aesthetics, functionality, safety, and manufacturability. Their success in preserving the essence of the TT's unique shape while creating a viable production car was a testament to their commitment to design excellence.

The result of this painstaking process was a production Audi TT that remained true to the original concept while meeting the requirements of a modern, road-legal sports car. It stood as a shining example of how, with careful consideration and skilled execution, the gap between concept and reality could be bridged without sacrificing the design's essence.

Section 2.5: Interior Design Evolution

The Audi TT's trailblazing exterior design was matched by an equally innovative interior, setting new standards for sports car cabins. The concept car's interior features were a testament to Audi's commitment to pushing boundaries and reimagining the driver's environment.

The concept TT's cabin was a futuristic masterpiece, featuring a minimalist design that emphasized clean lines and a driver-centric layout. It showcased a unique instrument cluster that combined analog and digital displays, foreshadowing the digital cockpits that would become commonplace in luxury vehicles years later. The center console was clutter-free, with essential controls seamlessly integrated into the dashboard's flowing design.

Translating this avant-garde interior design to production reality posed significant challenges. While some elements had to be toned down for practicality and cost considerations, Audi's designers worked tirelessly to preserve the essence of the concept's focused cabin. They focused on retaining the circular theme that echoed the car's exterior, incorporating it into air vents, door handles, and even the gear shift knob.

One of the most innovative elements retained from the concept was the aircraft-inspired cockpit feel. The production TT featured a driver-oriented dashboard that wrapped around the operator, creating a sense of control over a high-performance machine. This design choice not only enhanced the sporty feel but also improved ergonomics, placing all controls within easy reach of the driver.

The TT's interior designers paid meticulous attention to ergonomics and user experience. They crafted seats that provided both comfort for long drives and support for spirited cornering. The steering wheel, with its flat bottom, became an instant icon, enhancing both the car's sporty character and ease of ingress and egress.

Perhaps most notably, the TT's interior marked a significant departure from traditional Audi design language. While maintaining the brand's reputation for quality and attention to detail, the TT introduced a level of design flair previously unseen in Audi vehicles. The use of aluminum accents throughout the cabin, from the center console to the door panels, added a modern, technical feel that complemented the car's exterior.

The TT's interior also introduced novel material combinations. High-quality leather was juxtaposed with technical fabrics and metals, creating a tactile experience that was both luxurious and sporty. This blend of materials would go on to influence interior design across the automotive industry.

One of the most talked-about features of the TT's interior was its baseball-stitched leather seats, available as an option. This unique design element not only added visual interest but also provided a tangible link to the car's sporting heritage, evoking memories of leather driving gloves used in classic motorsports.

The attention to detail extended to even the smallest elements of the interior. The instrument cluster, for instance, featured a typeface that was both legible and stylish, further emphasizing the TT's design-forward approach.

Despite the need to adapt the concept's interior for production feasibility, the resulting cabin was a triumph of design. It captured the essence of the concept car while meeting the practical requirements of a production vehicle. The TT's interior set a new benchmark for sports-car cabins, proving that functionality and cutting-edge design can coexist harmoniously.

The impact of the TT's interior design was far-reaching. It influenced not only future Audi models but also set trends across the automotive industry. The emphasis on a driver-focused cockpit, the integration of digital and analog displays, and the use of high-quality materials in innovative ways became hallmarks of modern sports car interiors.

In essence, the evolution of the Audi TT's interior design from concept to production was a journey of calculated compromises and inspired solutions. The result was a cabin that was as brilliant as the car's exterior, cementing the TT's status as a design icon both inside and out.

Section 2.6: Final Production Design

As the Audi TT transitioned from concept to reality, the final production design emerged as a testament to the delicate balance between preserving the original vision and meeting practical manufacturing requirements. While the production model retained much of the concept's beautiful aesthetic, there were inevitably some differences between the show car and the vehicle that would roll off the assembly line.

One of the most notable changes was the slightly taller production TT than the concept. This adjustment was necessary to meet safety regulations and improve interior space, but designers worked tirelessly to ensure that the car's signature silhouette remained intact. The rounded greenhouse and pronounced wheel arches, hallmarks of the original design, were carefully preserved, maintaining the TT's unique visual identity.

The production TT's front fascia received subtle refinements, with the headlights adopting a more practical shape while retaining their look. The grille, while similar in overall design to the concept, was adapted to accommodate real-world cooling requirements and pedestrian safety standards. These changes, though necessary, were executed with such finesse that they hardly detracted from the car's overall aesthetic appeal.

What truly set the TT apart in the market was its unwavering commitment to clean, minimalist design. The absence of superfluous lines or unnecessary embellishments gave the TT a timeless quality that would help it stand out in a sea of over-styled competitors. The car's smooth, uninterrupted surfaces and perfect proportions created a rare sense of visual harmony in the automotive world of the late 1990s.

Audi offered the production TT in a carefully curated selection of colors and trims, each chosen to complement and enhance the car's unique form. From classic silvers and blacks to more daring choices like Nimbus Gray and Laser Red, each color option was selected to accentuate the TT's curves and surfaces. The interior trim options were carefully considered, ranging from leather to Alcantara, all designed to reinforce the car's premium positioning and avant-garde aesthetic.

The wheel designs for the production TT were a crucial element in completing the car's overall look. The standard 16-inch and optional 17-inch wheels were designed to fill the pronounced wheel arches perfectly, contributing to the car's planted, purposeful stance. The multi-spoke design of these wheels became almost as iconic as the car itself, perfectly complementing the TT's futuristic appearance.

In the final months before production launch, Audi's design team made a series of minute adjustments to ensure that every detail of the TT was perfect. From the precise curvature of the rear spoiler to the exact placement of the fuel filler cap, no element was too small to escape scrutiny. These final tweaks, while imperceptible to the casual observer, were crucial in elevating the TT from merely attractive to truly extraordinary.

The result of this meticulous design process was a production car that captured the essence of the original concept while meeting all the practical requirements of a mass-produced vehicle. The Audi TT, which finally reached showrooms in 1998, was a design triumph, proving that with enough dedication and skill, it was possible to translate a daring concept, almost unchanged, into a road-going reality. This achievement would cement the TT's place not only in Audi's lineup but also in the pantheon of automotive design classics.

Section 2.7: Design Legacy and Influence

The Audi TT's transformative design left an indelible mark on the automotive world, influencing not only Audi's future models but also reshaping the broader landscape of car design. From its inception, the TT's bold silhouette set it apart from its contemporaries, challenging conventional notions of what a sports car should look like.

The TT's influence on Audi's subsequent designs cannot be overstated. Its clean lines, minimalist approach, and perfect proportions became hallmarks of Audi's design language in the years that followed. The single-frame grille, which debuted on the TT, became a signature element across Audi's entire range. The TT's influence can be seen in models like the R8 supercar, which shares the TT's emphasis on geometric purity and visual balance.

Beyond Audi, the TT's impact rippled through the automotive industry. Its success proved that bold, avant-garde designs could be commercially viable, encouraging other manufacturers to take more risks with their styling. The TT's fusion of retro elements with futuristic touches inspired a wave of similarly nostalgic yet forward-looking designs across various brands.

The design community took notice of TT's design aesthetic. It garnered numerous awards and accolades, including the prestigious "Design of the Year" award from Automobile Magazine in 1999. The car's inclusion in various design museums and exhibitions further solidified its status as automotive art.

The TT's design played a crucial role in elevating it to an elite status. Its instantly recognizable shape became synonymous with progressive, cutting-edge design. Even those with little interest in cars could identify a TT, a testament to the power and memorability of its design. This visual impact contributed significantly to the TT's enduring popularity and cult following.

The lessons learned from the TT's design process were numerous and far-reaching. It demonstrated the importance of staying true to a strong initial concept, even amid production challenges. The TT's development also highlighted the value of cross-disciplinary collaboration, with designers and engineers working closely to bring the vision to life without compromising its essence.

Perhaps most importantly, the TT's journey from sketch to showroom floor illustrated the transformative power of innovative design. It showed how a single model could redefine a brand's image and set new standards for an entire industry. The TT proved that cars could be more than mere transportation; they could be rolling sculptures, captivating the imagination and stirring the soul.

The Audi TT: A Journey Through Design and Innovation

As we reflect on the TT's design legacy, it's clear that its influence extends far beyond its sales figures or performance statistics. It changed perceptions, pushed boundaries, and inspired a generation of designers and car enthusiasts. The Audi TT stands as a testament to the enduring power of bold, visionary design in the automotive world.

Chapter 3: Engineering Excellence: The TT's Technical Foundations

Section 3.1: The Platform Revolution

The Audi TT's journey to becoming an automotive icon began with an innovative approach to vehicle architecture: the PQ34 platform. This shared platform strategy, pioneered by the Volkswagen Group, laid the foundation for the TT's unique blend of performance, style, and efficiency.

The PQ34 platform, also known as the A4 platform, was a marvel of engineering versatility. Initially developed for compact cars like the Volkswagen Golf and Audi A3, it demonstrated the Volkswagen Group's commitment to streamlining production while maintaining distinct brand identities. The platform's adaptability allowed it to underpin a diverse range of vehicles, from family hatchbacks to high-performance sports cars like the TT.

Adapting this platform for a sports car presented both challenges and opportunities for Audi's engineers. The team faced the challenge of transforming a foundation designed for everyday vehicles into one capable of meeting the dynamic demands of a sports coupe.

This required extensive modifications, including significant structural reinforcements to enhance rigidity and precise adjustments to weight distribution.

One of the most innovative aspects of the TT's development was the extensive use of aluminum through Audi Space Frame (ASF) technology. This lightweight construction method allowed the TT to shed excess weight without compromising structural integrity. The result was a sports car that weighed just 1,240 kg for the base model, significantly lighter than many of its competitors. This weight advantage translated directly into improved acceleration, handling, and fuel efficiency.

Aerodynamics played a crucial role in shaping the TT's performance characteristics. The car's iconic rounded silhouette wasn't just a design statement; it was the result of extensive wind tunnel testing and refinement. Engineers meticulously sculpted every curve and contour to minimize drag while maintaining stability at high speeds. The result was a drag coefficient of just 0.32, an impressive figure for a sports car of its time.

To achieve the agility expected of an actual sports car, Audi's chassis engineers worked tirelessly to tune the TT's suspension. They opted for a MacPherson strut front suspension and a multi-link rear setup, a combination that provided an ideal balance of comfort and responsiveness. The suspension geometry was carefully calibrated to minimize body roll and maximize grip, allowing the TT to corner with precision and confidence.

This chassis tuning set the TT apart from other Audi models, giving it a distinctly sportier feel. While cars like the A4 sedan prioritized comfort and refinement, the TT's suspension was tuned to deliver immediate driver feedback and respond quickly to inputs. This setup allowed the TT to feel nimble and engaging on twisty roads while still maintaining the composure expected of an Audi during high-speed autobahn cruising.

The platform revolution didn't stop at the chassis. Audi's engineers also focused on optimizing the car's weight balance. By carefully positioning components and using lightweight materials where possible, they achieved a near-ideal weight distribution. This balanced approach contributed significantly to the TT's neutral handling, enabling it to change direction with ease and maintain stability during high-speed maneuvers.

In essence, the PQ34 platform provided Audi with a versatile canvas on which to paint its sports car masterpiece. By adapting this foundation with innovative materials, aerodynamic refinements, and precise suspension tuning, Audi created a sports car that was both technically advanced and emotionally engaging. The TT's platform wasn't just a base to build upon; it was a key element in defining the car's character, setting the stage for a driving experience that would captivate enthusiasts for generations to come.

Section 3.2: Powertrain Innovation

The heart of any sports car lies in its powertrain, and the Audi TT is no exception. From its inception, the TT has been a showcase of Audi's engineering prowess, offering a range of engines that combine performance, efficiency, and character.

At the core of the TT's powertrain lineup is the renowned 1.8T engine. This turbocharged four-cylinder powerplant became synonymous with the TT, offering a compelling blend of performance and efficiency. In its initial form, the 1.8T produced a respectable 180 horsepower, but Audi engineers quickly realized its potential for more. Subsequent iterations saw power outputs increase to 225 horsepower and beyond, with some variants approaching 250 horsepower. This continuous development demonstrated Audi's commitment to evolving the TT's performance capabilities throughout its lifecycle.

The 1.8T wasn't just about raw power figures; it was the engine's character that truly set it apart. With a broad torque curve that delivered peak twist from as low as 1,800 rpm, the TT offered effortless acceleration and responsiveness that belied its relatively small displacement. This characteristic made the TT equally at home navigating city streets or carving through mountain passes.

As the TT evolved, so did its engine options. Later generations introduced larger-displacement engines, including a potent 3.2-liter V6 that not only increased power but also added an exhaust note that enthusiasts craved. This expansion of the engine range allowed Audi to cater to a broader array of customers, from those seeking efficient daily drivers to hardcore performance enthusiasts.

One of the TT's most defining features is its Quattro all-wheel-drive system. While Quattro was already a hallmark of Audi's performance models, adapting it for use in a compact sports car presented unique challenges and opportunities. The TT's Quattro system was tuned not only for all-weather traction but also to enhance the car's handling dynamics. By continuously varying torque distribution between the front and rear axles, the system could alter the TT's handling characteristics on the fly, providing a level of adaptability that was rare in its class.

In normal driving conditions, the Quattro system typically sends about 90% of the power to the front wheels for efficiency. However, it could instantly redirect up to 100% of available torque to the rear wheels when needed, such as during spirited cornering or in low-traction situations. This capability gave the TT a level of sure-footedness that set it apart from its primarily rear-wheel-drive competitors, making it a favorite among enthusiasts who valued all-weather performance.

The TT's transmission options were equally well-considered. Early models offered a choice between a six-speed manual transmission for purists and a six-speed automatic for those prioritizing convenience. The manual gearbox was praised for its precise shifts and short throw, enhancing the driver's connection to the car. As technology advanced, later TT models introduced dual-clutch transmissions, offering lightning-fast gear changes that could outperform even the most skilled manual shifters in acceleration.

Audi's engineers paid particular attention to the TT's exhaust system, recognizing that sound plays a crucial role in the sports car experience. The challenge was to create an exhaust note that was sporty and engaging without being overly aggressive or droning on long trips. Through careful tuning of the exhaust system's geometry and the use of variable valves, they achieved a sound that was subdued during everyday driving but came alive under hard acceleration, adding an auditory thrill to the driving experience.

Managing heat in a compact sports car presents unique challenges, and the TT's cooling system is a testament to Audi's innovative approach. High-performance variants, in particular, required advanced cooling solutions to maintain optimal operating temperatures under demanding conditions. This led to the development of more efficient radiators, carefully designed air intakes, and in some cases, additional oil coolers for the engine and transmission.

The powertrain innovations in the Audi TT weren't just about raw performance; they were about creating a holistic driving experience that balanced power, efficiency, and character. From the responsive turbocharged engines to the adaptable Quattro system, from the engaging transmissions to the carefully tuned exhaust note, every aspect of the TT's powertrain was engineered to enhance the connection between car and driver. This commitment to powertrain excellence has been a key factor in the TT's enduring appeal, ensuring its place as a benchmark in the sports car segment.

Section 3.3: Electronic Integration

As automotive technology advanced rapidly in the late 1990s and early 2000s, the Audi TT embraced cutting-edge electronic systems to enhance performance, safety, and driver experience. This section explores the key electronic innovations that set the TT apart from its competitors and contributed to its reputation as a technologically advanced sports car.

The introduction of drive-by-wire technology marked a significant leap forward in the TT's electronic integration. This system replaced traditional mechanical linkages between the driver's inputs and the car's responses with electronic sensors and actuators. In the TT, this technology was most notably applied to the throttle control. Instead of a cable connecting the accelerator pedal to the engine, sensors detected pedal movement and relayed this information to the engine control unit, which then adjusted the throttle accordingly. This enabled more precise control over engine response and allowed advanced traction control and cruise control to be implemented more effectively. Drivers noticed improved throttle response and a more connected feel to the car's acceleration.

Advanced traction control systems were another area where the TT showcased its electronic prowess. These systems were designed to enhance stability without compromising the car's sporty character. Unlike earlier, more intrusive systems that simply cut power at the first sign of wheel slip, the TT's traction control used sophisticated algorithms to monitor wheel speed, yaw rate, and steering angle. This allowed the system to intervene more subtly, applying brake pressure to individual wheels or modulating engine power to maintain traction while still allowing for spirited driving. In wet or slippery conditions, this system proved invaluable, giving drivers confidence without dulling the TT's responsive handling.

The Audi TT's brake system also benefited from electronic innovations. High-performance braking was a must for a sports car, and the TT delivered with an advanced anti-lock braking system (ABS) and electronic brake-force distribution (EBD). These systems worked in tandem to ensure optimal braking performance under various conditions. The ABS prevented wheel lock-up during hard braking, allowing the driver to maintain steering control, while the EBD dynamically adjusted brake force between the front and rear wheels based on load and traction conditions. This resulted in shorter stopping distances and improved stability during emergency braking maneuvers. Moreover, the brake system was designed to resist fade even under the demanding conditions of track driving, a testament to the TT's performance credentials.

While the TT was primarily focused on driving dynamics, it didn't ignore the growing importance of infotainment and connectivity. The challenge was to integrate modern technology without detracting from the driver-focused cockpit design. Early TT models featured relatively simple audio systems, but as the model evolved, so did its infotainment offerings. Later generations introduced more advanced systems, including navigation, Bluetooth connectivity, and Audi's Virtual Cockpit in the third generation. This digital instrument cluster enabled a customizable display of vehicle information, navigation, and entertainment options, right in the driver's line of sight, maintaining the TT's ethos of putting the driver first.

Lighting technology was another area where the TT showcased electronic innovation. The first-generation TT introduced xenon headlights as an option, providing brighter, more focused illumination than traditional halogen bulbs. As LED technology advanced, it was incorporated into the TT's lighting design, both for functionality and as a styling element. LED daytime running lights became a signature feature, enhancing visibility and giving the TT an unmistakable presence on the road. In later models, full LED headlights offered even better illumination, faster response times, and lower energy consumption. Some variants even featured adaptive lighting systems that could adjust the beam pattern based on speed and steering input, improving visibility on winding roads.

The electronic integration in the Audi TT went beyond individual systems; it was about creating a cohesive, responsive driving experience. The car's various electronic systems communicated via a sophisticated Controller Area Network (CAN), enabling rapid information exchange and coordinated responses to driver inputs and changing road conditions. This integration was key to the TT's ability to deliver both sports car performance and everyday usability.

In conclusion, the electronic integration in the Audi TT represented a careful balance between cutting-edge technology and pure driving engagement. From drive-by-wire systems to advanced traction control, high-performance brakes, and innovative lighting, these electronic systems enhanced the TT's performance, safety, and appeal. They demonstrated Audi's commitment to embracing new technologies while maintaining the essential character of a driver's car. As we'll see in later chapters, this foundation of electronic innovation would continue to evolve, keeping the TT at the forefront of automotive technology throughout its lifetime.

Section 3.4: Material Science

The Audi TT's engineering excellence extends beyond its mechanical components and into the realm of material science. This section explores how Audi's innovative use of materials contributes to the TT's performance, safety, and overall appeal. At the core of the TT's construction is the strategic use of high-strength steel. This advanced material allows Audi to create a body structure that is both lightweight and incredibly rigid. The high-strength steel is used in critical areas of the chassis, providing excellent crash protection while minimizing weight.

This careful balance results in improved handling characteristics and fuel efficiency without compromising safety. In crash tests, the TT consistently outperforms expectations, demonstrating the effectiveness of this material choice.

Complementing the steel structure are composite materials used in various body panels. These advanced composites offer significant weight savings compared to traditional steel panels while maintaining excellent durability. For instance, the TT's hood and some exterior panels are made from these materials, contributing to the car's overall weight reduction. This approach not only enhances performance but also improves fuel efficiency, making the TT more environmentally friendly without sacrificing its sporty character.

Audi's engineers have also pioneered advanced adhesives and joining techniques in the TT's construction. Moving beyond traditional welding, these methods enable more precise, stronger connections between different materials. This innovative approach significantly improves the TT's structural rigidity, enhancing both safety and handling. The result is a car that feels solid and well-built, with minimal flex even under high-stress driving conditions.

The material science innovations extend to the TT's interior as well. The selection of interior materials strikes a delicate balance among luxury, durability, and weight. High-quality leathers and synthetic materials are chosen not just for their aesthetic appeal but also for their ability to withstand wear and tear while keeping the overall weight down. A standout feature of the TT's interior is its extensive use of aluminum accents. These not only contribute to the car's look but also help reduce weight and improve heat dissipation.

The car's shape is protected and enhanced by advanced paint formulations that offer superior resistance to chips, scratches, and environmental factors. Audi has developed special color options for the TT, including unique metallic and pearl effect finishes that accentuate the car's curves and lines. These paints are not just about aesthetics; they also provide an additional layer of protection to the body, contributing to the car's longevity.

The material science behind the Audi TT is a testament to the company's commitment to innovation. By carefully selecting and implementing advanced materials throughout the car, Audi has created a sports car that is not only high-performing and safe but also efficient and durable. This holistic approach to material use is a key factor in the TT's enduring appeal and its ability to stand out in a competitive market.

Section 3.5: Performance Enhancements

The Audi TT's reputation as a high-performance sports car is built on a foundation of cutting-edge engineering and relentless pursuit of automotive excellence. This section delves into the key performance enhancements that have elevated the TT from a stylish coupe to a true driver's car.

At the heart of the TT's performance lies its advanced turbocharging system. Audi's engineers have continuously refined the turbocharging technology, pushing the boundaries of what's possible from a compact engine. The evolution of turbo design in the TT has been nothing short of amazing.

Early models suffered from noticeable turbo lag, a common issue in turbocharged engines of the era. However, with each subsequent generation, Audi has significantly reduced this lag, creating a more responsive and exhilarating driving experience. The latest TT models feature turbochargers that spool up almost instantly, delivering immediate power that belies the engine's modest displacement. This advancement is particularly evident when comparing the throttle response of a first-generation TT to its modern counterpart; the difference is night and day.

Complementing the turbocharger is an intelligently designed intercooler system. The intercooler plays a crucial role in maximizing engine performance by cooling the compressed air from the turbocharger before it enters the engine. Audi's engineers have continuously refined the intercooler's design and placement to optimize its efficiency.

In the TT, the intercooler's location has been carefully chosen to maximize airflow and minimize pressure drop. This attention to detail results in denser air entering the engine, allowing for more fuel to be burned and, consequently, more power to be produced. The efficiency gains from these intercooler improvements are particularly noticeable in high-performance variants like the TT RS, where every horsepower counts.

Variable valve timing is another key technology that Audi has implemented to enhance the TT's performance. This system allows optimal valve operation across the entire rev range, effectively giving the TT two engines in one: one optimized for low-end torque and fuel efficiency, and another for high-end power. The result is a car that's both docile in city driving and explosive on the open road. For instance, the latest TT models equipped with this technology exhibit a remarkably flat torque curve, providing strong acceleration from just above idle to the redline.

The TT's quattro all-wheel-drive system has always been a defining feature, but it's the differential technology within this system that truly sets it apart. Audi has progressed from open differentials in early models to sophisticated limited-slip and electronic differentials in later generations. These advanced differentials can instantly vary the torque split between the front and rear axles, and even between the left and right wheels.

This technology shines in challenging driving conditions, allowing the TT to put its power down effectively regardless of road surface or cornering forces. The result is a car that feels planted and confident in all situations, from hairpin mountain roads to rain-soaked highways.

Finally, Audi's engineers have paid meticulous attention to the TT's weight distribution. Despite being based on a front-wheel-drive platform, the TT achieves a near-perfect 50:50 front-to-rear weight balance. This has been accomplished by carefully placing components throughout the car. For example, the battery is often located in the rear of the vehicle to offset the weight of the front-mounted engine. Even the placement of smaller components, such as the windshield washer fluid reservoir, has been considered in this balancing act. The result is a car that feels incredibly neutral and balanced in corners, responding predictably to driver inputs and maintaining composure even when pushed to its limits.

These performance enhancements work in concert to create a driving experience that is uniquely Audi TT. From the instant throttle response provided by advanced turbocharging to the confident cornering enabled by sophisticated differentials and ideal weight distribution, every aspect of the TT's performance has been honed to perfection. It's this attention to detail and commitment to continuous improvement that have kept the TT at the forefront of the sports car market for over two decades.

Section 3.6: Aerodynamics and Design Synergy

The Audi TT's shape isn't just a triumph of aesthetics; it's a masterclass in the fusion of form and function. From its inception, the TT's designers and engineers worked in tandem to create a vehicle that would not only turn heads but also slice through the air with remarkable efficiency. At the heart of the TT's aerodynamic prowess is its unique rounded silhouette. This shape wasn't chosen merely for its visual appeal; it was meticulously crafted to minimize air resistance.

The smooth, unbroken lines from the front bumper to the rear diffuser create a seamless flow, allowing air to glide over the car's surface with minimal turbulence. This reduces drag and improves fuel efficiency, a crucial factor in the TT's overall performance.

The TT's front fascia is a study in purposeful design. The signature Audi grille, while bold, is carefully shaped to channel air into the engine bay for cooling. Flanking the grille, the headlights are smoothly integrated into the body, avoiding unnecessary protrusions that could disrupt airflow. Even the side mirrors, often an afterthought in car design, are sculpted to reduce wind noise and drag.

Moving along the side of the vehicle, the TT's wheel arches are subtly flared, not just for aesthetic appeal but to accommodate wider tires without creating additional drag. The smooth transition from the hood to the windshield, and the gentle slope of the roof to the rear create a continuous arc through which air can flow uninterrupted.

One of the most innovative aerodynamic features of the TT is its retractable rear spoiler. In earlier models, this spoiler would automatically deploy at higher speeds, providing additional downforce to improve stability. Later iterations refined this feature, optimizing the spoiler's deployment speed and angle based on extensive wind-tunnel testing and real-world data. This active aerodynamic element allows the TT to maintain its clean lines at lower speeds while still benefiting from improved high-speed stability.

The rear of the TT is equally well-considered from an aerodynamic standpoint. The sloping rear window and short deck lid help to reduce the low-pressure area behind the car, minimizing drag. The diffuser integrated into the rear bumper isn't just for show; it helps to manage airflow underneath the car, reducing lift and improving stability.

Even the TT's underbody has been given careful attention. Smooth panels help guide airflow, reducing turbulence and the associated drag. This attention to detail extends to components like the exhaust system and rear suspension, which are designed to minimize disruption to airflow.

The result of this meticulous attention to aerodynamics is a car that not only looks stunning but also performs exceptionally well. The TT's low drag coefficient allows it to achieve impressive fuel efficiency for a sports car, while also contributing to its high-speed stability and performance.

The most remarkable aspect of the TT's aerodynamics is how seamlessly they're integrated into the overall design. Unlike some high-performance vehicles that sacrifice elegance for function, the TT achieves impressive aerodynamic performance without resorting to obvious spoilers, vents, or other add-ons. It's a testament to Audi's designers and engineers that they created a car that's both beautiful and aerodynamically efficient.

This synergy between design and aerodynamics is a key part of what makes the TT special. It's not just a sports car that performs well; it's a work of automotive art, carefully sculpted to slice through the air with grace and efficiency. This holistic approach to design and engineering is a hallmark of the TT, setting it apart in the competitive world of sports cars and contributing significantly to its enduring appeal among enthusiasts and casual observers alike.

Section 3.7: The Future of TT Engineering

As we look towards the horizon of automotive engineering, the Audi TT continues to evolve, pushing the boundaries of what's possible in a compact sports car. The future of TT engineering is a tantalizing blend of cutting-edge technology, sustainable practices, and the timeless pursuit of driving pleasure.

At the forefront of this evolution is electrification. Audi has already made significant strides in electric vehicle technology with models like the e-tron, and it's only a matter of time before this expertise finds its way into the TT lineup. Imagine a fully electric TT, offering instant torque and a low center of gravity thanks to floor-mounted batteries. This isn't just a flight of fancy; it's a very real possibility that could redefine the TT's performance characteristics while maintaining its commitment to environmental responsibility.

But electrification isn't the only path forward. Audi is also exploring synthetic fuels to reduce the carbon footprint of internal combustion engines. These e-fuels, as they're often called, could allow future TT models to maintain the visceral thrill of a traditional engine while significantly reducing emissions. It's a best-of-both-worlds scenario that could appeal to purists and environmentally conscious drivers alike.

Advanced materials will also play a crucial role in TT's future. We're likely to see increased use of carbon fiber and other composites, not just in body panels but also in structural components. These materials offer exceptional strength-to-weight ratios, potentially making future TT models even lighter and more agile than their predecessors.

Artificial intelligence and machine learning are set to revolutionize how the TT interacts with its driver. Adaptive suspension systems that learn your driving style and adjust accordingly, predictive maintenance alerts that keep your TT in peak condition, and advanced driver assistance features that enhance safety without diluting the driving experience are all on the horizon.

Connectivity will also play a larger role in future TT models. Expect seamless integration with smart devices, over-the-air updates that enhance performance and add features, and augmented reality displays that overlay navigation and vehicle information onto the real world.

Perhaps most excitingly, Audi is likely to continue pushing the boundaries of aerodynamics. Active aero elements that adjust in real-time to optimize downforce and reduce drag could take the TT's already impressive handling to new heights. Coupled with advanced tire technology and even more sophisticated all-wheel-drive systems, future TT models could offer levels of grip and control that would make current sports cars blush.

Of course, with all these advancements, Audi faces the challenge of maintaining what makes a TT a TT. The design, the driver-focused interior, and the perfect balance of everyday usability and sports car thrills are all integral to the TT's identity. As such, we can expect future engineering efforts to enhance these core attributes rather than replace them.

In conclusion, the future of TT engineering is bright, filled with possibilities that seemed like science fiction not too long ago. Whether powered by electricity, synthetic fuels, or some as-yet-undiscovered energy source, the TT of tomorrow will continue to embody Audi's commitment to progress through technology. It will be lighter, faster, more efficient, and more connected than ever before, yet it will remain unmistakably a TT, a sports car that captures the imagination and stirs the soul.

The Audi TT: A Journey Through Design and Innovation

Chapter 4: The First Generation: Revolutionizing Sports Car Design

Section 4.1: The Unveiling of a Legend

The automotive world stood still on September 10, 1998, as Audi pulled the covers off the production TT at the Frankfurt Motor Show. The crowd's collective gasp was audible as they witnessed a car that looked remarkably close to the concept shown three years earlier. This was no ordinary unveiling; it was the birth of a legend that would redefine sports car design for years to come.

Unlike many concept-to-production transitions that often disappoint enthusiasts, the TT retained about 80% of its concept design. This feat impressed both critics and enthusiasts alike, setting a new standard for concept car fidelity in production models. The TT's debut was a testament to Audi's commitment to pushing the boundaries of automotive design and engineering.

The immediate response from the public and automotive press was nothing short of euphoric. Car magazines worldwide hailed the TT as a design masterpiece, with one prominent reviewer calling it "the Bauhaus on wheels." This comparison to the influential German art school known for its modernist approach was fitting, as the TT's clean lines and minimalist aesthetic embodied the Bauhaus principles of form following function.

The TT's introduction had a profound impact on Audi's brand image. Almost overnight, it elevated Audi's status in the sports car segment, positioning the brand as a serious contender against established players like Porsche and BMW. The TT became more than just a new model; it was a statement of intent from Audi, showcasing their ability to produce cutting-edge, desirable sports cars that could compete with the best in the world.

Audi's marketing team deserves credit for building unprecedented anticipation for the TT's launch. Their clever "TT-teaser" campaign in automotive magazines tantalized potential buyers and enthusiasts alike. These cryptic ads showcased only parts of the car's unique silhouette, sparking speculation and excitement in the months leading up to the Frankfurt reveal. This strategy proved highly effective, ensuring that all eyes were on Audi when the TT finally made its debut.

The unveiling of the first-generation Audi TT was more than just the launch of a new sports car; it was a pivotal moment in automotive history. It challenged conventional design norms, raised the bar for concept-to-production fidelity, and reshaped Audi's image in the competitive sports car market. The legend of the TT was born on that September day in Frankfurt, and its influence would be felt throughout the automotive industry for years to come.

Section 4.2: Revolutionary Design Elements

The first-generation Audi TT's design was nothing short of groundbreaking, introducing a host of innovative elements that would influence automotive aesthetics for years to come. At the heart of its appeal was the silhouette, a perfect balance of proportions that created an instantly recognizable profile. The TT's unique roofline, with its gentle arc sweeping from the A-pillar to the rear, combined with the pronounced wheel arches and short overhangs, resulted in a shape that was both bold and harmonious. This unique silhouette not only set the TT apart from its contemporaries but also became a design touchstone, often imitated but rarely equaled.

One of the most innovative aspects of the TT's design was its pioneering use of aluminum. Audi's extensive use of aluminum not only reduced the vehicle's overall weight but also enabled crisper body lines and tighter panel fits than traditional steel construction. This approach resulted in a car that looked as if it had been milled from a solid block of metal, with a level of precision and quality that was unprecedented in its class. The use of aluminum also contributed to the TT's performance and efficiency, setting a new standard for sports car engineering.

The TT's interior design philosophy was as innovative as its exterior. Adopting an aircraft-inspired design language, the cabin featured a circular theme that was both aesthetically pleasing and functionally intuitive. From the round air vents to the circular instrument cluster, every element reinforced this cohesive design approach. The minimalist layout, free from unnecessary ornamentation, created a sense of purpose and focus that perfectly complemented the car's sporting intentions. This clean, driver-centric design set new standards for sports car interiors, influencing cabin layouts across the industry.

Lighting innovations were another area where the TT pushed boundaries. The use of clear-lens headlights was relatively new at the time, and the TT's implementation was particularly striking. The rear light clusters, with their unique shape and arrangement, became a signature element of the design. These lighting elements not only enhanced the car's visual appeal but also improved visibility and safety. The impact of the TT's lighting design can be seen in countless vehicles that followed, as manufacturers sought to emulate its crisp, modern aesthetic.

Audi offered the TT in a range of bold colors and high-quality materials, enabling extensive personalization. One of the most memorable options was the Baseball Optic leather seats, which featured a unique stitching pattern reminiscent of a baseball glove. This attention to detail in color and trim options demonstrated Audi's commitment to creating a truly special vehicle. The combination of avant-garde design with high-quality, tactile materials elevated the TT above its competitors, positioning it as a premium product in the sports car market.

The TT's trailblazing design elements worked in harmony to create a car that was more than the sum of its parts. Its blend of innovative construction techniques, bold styling choices, and meticulous attention to detail resulted in a vehicle that looked like nothing else on the road. The first-generation TT didn't just introduce new design ideas; it challenged the very notion of what a sports car could be, paving the way for a new era of automotive design that prioritized form alongside function. Its influence extended far beyond the realm of sports cars, inspiring designers across various industries and cementing its status as a true design icon of the late 20th and early 21st centuries.

Section 4.3: Technical Specifications and Powertrains

The first-generation Audi TT's trailblazing design was matched by an equally impressive array of technical specifications and powertrain options. At its launch, the TT offered a choice of engines that balanced performance with efficiency, catering to a range of driving preferences.

The TT debuted with a 1.8-liter turbocharged four-cylinder engine, available in two power outputs. The base model produced a respectable 180 horsepower, while the more potent variant 0delivered 225 horsepower. These engines were known for their responsive nature and strong mid-range torque, characteristics that perfectly suited the TT's sporty persona. In 2003, Audi expanded the engine lineup with the introduction of a 3.2-liter VR6 engine. This naturally aspirated powerplant brought a new dimension to the TT's performance, offering 250 horsepower and an exhaust note that enthusiasts adored.

One of the TT's key technical advantages was Audi's renowned quattro all-wheel-drive system. Available on higher-powered models, the quattro system provided exceptional traction and handling, setting the TT apart from many of its rear-wheel-drive competitors. This feature not only enhanced the car's performance capabilities but also improved its all-weather usability, making it a practical choice for drivers in various climates.

Transmission options for the TT were equally impressive. Purists appreciated the precise 6-speed manual gearbox, which offered crisp shifts and an engaging driving experience. However, it was the innovative dual-clutch S tronic transmission that truly showcased Audi's technical prowess. This transmission offered lightning-fast shifts and improved fuel efficiency, providing a perfect blend of performance and convenience. The S tronic was particularly well-suited to the TT's character, allowing drivers to enjoy both relaxed cruising and spirited driving with equal ease.

The TT's suspension setup was carefully tuned to deliver a balance between comfort and sportiness. The front suspension utilized MacPherson struts, while the rear employed a sophisticated multi-link arrangement. This combination provided sharp handling and good road feel without compromising ride quality, a crucial factor for a car that was as much about style as it was about performance. The suspension geometry was later revised as part of Audi's response to early stability concerns, further improving the TT's high-speed stability and overall dynamics.

In terms of performance figures, the TT impressed across its range. The top-spec TT quattro Sport could sprint from 0-60 mph in just 5.7 seconds, an impressive feat for its time. Even the base models offered spirited performance, with the 180 hp variant reaching 60 mph in about 7.5 seconds. Top speeds varied across the range, with the most powerful models capable of reaching electronically limited top speeds of 155 mph.

The TT's relatively lightweight, achieved through aluminum construction, played a significant role in its performance. This lightweight approach not only improved acceleration and handling but also contributed to better fuel efficiency compared to many of its competitors.

As the first generation progressed, Audi continued to refine and enhance the TT's technical specifications. Updates included improvements to engine efficiency, transmission software upgrades, and subtle chassis tweaks, all aimed at keeping the TT competitive in an increasingly crowded sports car market.

The combination of these technical elements, powerful engines, advanced all-wheel drive, innovative transmissions, and a well-tuned chassis created a driving experience that was both exhilarating and accessible. The TT offered performance that could satisfy enthusiasts while remaining manageable for less experienced drivers, a balance that contributed significantly to its broad appeal and commercial success.

In essence, the technical specifications and powertrains of the first-generation Audi TT were a perfect complement to its thoughtful design. They provided the substance to match the style, ensuring that the TT was not just a pretty face in the sports car world, but a genuine performer capable of delivering an engaging driving experience.

Section 4.4: Safety Innovations and Challenges

The first-generation Audi TT not only revolutionized sports car design but also pushed the boundaries of automotive safety. However, this journey was not without its challenges, as the car's design initially presented some unexpected safety concerns.

Shortly after its launch, reports began to emerge of stability issues at high speeds. Some drivers experienced unexpected oversteer, particularly during sudden lane changes or evasive maneuvers at highway speeds. This raised concerns about the TT's safety and threatened to tarnish its otherwise stellar reputation.

Audi, recognizing the gravity of the situation, took swift and decisive action to address these concerns. The company's response was twofold, combining electronic innovations with physical modifications to enhance the TT's stability and safety.

The most significant development was the implementation of the Electronic Stability Program (ESP) as standard equipment on all TTs. This advanced system continuously monitored the car's movement and could intervene by applying individual brakes or modulating engine power to maintain stability. The introduction of ESP significantly improved the TT's high-speed stability and overall safety profile.

In addition to the electronic aids, Audi made several physical modifications to the car. A rear spoiler was added to increase downforce at high speeds, improving stability. The suspension system was also recalibrated, with adjustments to spring rates and damper settings to improve handling predictability.

These modifications were not merely Band-Aid solutions but comprehensive engineering changes that addressed the root causes of the stability issues. Audi offered to retrofit these improvements to all existing TTs, demonstrating their commitment to customer safety and satisfaction.

Despite the initial challenges, the TT's safety features were impressive for its time. The car featured a robust safety cage designed to protect occupants in the event of a collision. Strategically placed crumple zones were engineered to absorb impact energy, while high-strength steel in critical areas enhanced structural integrity.

The TT also incorporated advanced passive safety systems, including dual front airbags and side-impact airbags for both driver and passenger. The seats were designed with integrated headrests to reduce the risk of whiplash in rear-end collisions.

Audi's handling of the stability issues demonstrated the company's commitment to safety and helped maintain consumer confidence in the TT. The swift and comprehensive response, which included a voluntary recall and free upgrades for existing owners, was praised by safety advocates and helped preserve the TT's reputation.

This episode also had broader implications for the automotive industry. It highlighted the importance of thorough real-world testing for vehicles with unconventional designs and led to more stringent stability requirements for sports cars across the board.

In the years following the stability enhancements, the TT earned solid safety ratings in various crash tests conducted by independent organizations. These results helped reassure potential buyers and contributed to the car's continued success in the marketplace.

The safety journey of the first-generation Audi TT serves as a case study in how automakers can effectively respond to unexpected challenges. It demonstrates that with the right combination of technological innovation and engineering expertise, even significant safety concerns can be overcome, resulting in a safer, more capable vehicle.

As the TT evolved, these early safety innovations became the foundation for even more advanced systems in subsequent generations, ensuring that the TT remained not just a style icon but also a leader in sports-car safety.

Section 4.5: Special Editions and Variants

The first-generation Audi TT's popularity and unique design led to several special editions and variants, each offering distinctive features and experiences for enthusiasts. One of the most significant additions to the TT lineup was the TT Roadster, introduced in 1999. This open-top variant retained the coupe's lines while offering the thrill of open-air driving. The Roadster's well-designed folding soft top integrated seamlessly into the TT's overall aesthetic, allowing drivers to enjoy the elements without compromising the car's visual appeal.

For those seeking the ultimate expression of the first-generation TT's potential, Audi introduced the limited-edition TT quattro Sport. This variant represented the pinnacle of performance for the model, featuring increased power output and reduced weight. The quattro Sport's enhancements included a more potent engine tune, lighter components, and a more focused suspension setup, resulting in a sharper, more responsive driving experience that truly showcased the TT's sporting credentials.

In 2003, Audi expanded the TT's powertrain options with the introduction of the TT 3.2 VR6. This addition of the 3.2-liter VR6 engine gave the TT a significant power boost, enhancing its grand touring capabilities. The smooth, sonorous V6 engine provided a different character to the TT, appealing to drivers who desired more refined performance and a more relaxed cruising ability.

Audi also catered to specific markets with regional special editions. In the United Kingdom, for example, Audi offered the TT 'Nimbus' edition. This variant featured unique paint and interior combinations not available on standard models, allowing buyers to own a more exclusive version of the already TT. These regional editions often became highly sought after by collectors due to their rarity and unique specifications.

While not directly available as a production model, the TT-R DTM race car significantly influenced several sport-oriented options and limited editions of the road-going TT. The motorsport pedigree gained from the TT's racing exploits trickled down to special road versions, often featuring enhanced aerodynamics, more powerful engines, or track-focused suspension setups.

These special editions and variants not only broadened the TT's appeal but also demonstrated the TT's design versatility. From the wind-in-your-hair freedom of the Roadster to the track-inspired performance of the quattro Sport, each variant maintained the essence of the TT while offering a unique driving experience. The introduction of these models throughout the first generation's lifecycle helped keep interest in the TT. It solidified its position as a versatile sports car capable of satisfying a wide range of enthusiasts.

The diversity of TT variants also showcased Audi's commitment to innovation and its ability to extract the platform's full potential. Whether catering to sun-seekers, performance enthusiasts, or those desiring a more powerful grand tourer, Audi ensured that there was a TT to suit various tastes and driving styles. This approach not only boosted sales but also contributed to the TT's enduring legacy, creating a devoted following that appreciated the model's ability to evolve while maintaining its core identity.

Section 4.6: The TT's Impact on Audi's Lineup

The introduction of the first-generation Audi TT had a profound and lasting impact on Audi's entire model range, influencing everything from design language to technology implementation. The TT's bold and innovative approach to sports car design didn't just revolutionize its segment; it sent ripples through the automotive industry and transformed Audi's brand identity.

The TT's design language began to permeate other Audi models shortly after its debut. The clean lines, minimalist aesthetic, and attention to detail that characterized the TT started to appear across Audi's range. Models from the compact A3 to the flagship A8 began incorporating elements of the TT's design philosophy. This influence was particularly noticeable in the interiors of subsequent Audi vehicles, where the TT's driver-focused cockpit and high-quality materials set a new standard for the brand.

Beyond aesthetics, the TT also served as a technological pioneer for Audi. The aluminum space frame construction techniques developed for the TT were later applied to larger Audi models, improving their performance and efficiency. This lightweight construction method became a hallmark of Audi's engineering prowess, allowing the brand to create vehicles that were both powerful and fuel-efficient. The TT's quattro all-wheel-drive system, while not new to Audi, was refined for the sports car, and these improvements trickled down to other models in the lineup.

Perhaps even more significant than its tangible influences was the TT's effect on Audi's brand image. The TT's success and critical acclaim boosted Audi's reputation as a manufacturer of desirable, design-led vehicles. This halo effect helped drive sales across the entire model range, as customers began to view Audi in a new light. The brand was no longer seen as just a maker of reliable German cars, but as a purveyor of stylish, technologically advanced vehicles that offered both performance and prestige.

This shift in perception had a profound impact on Audi's marketing strategy. The TT's success allowed Audi to position itself as a more emotive, design-focused brand. Marketing materials began to emphasize not just the technical specifications of Audi vehicles, but also their aesthetic appeal and the emotional connection they could create with drivers. This strategy continues to this day, with Audi consistently emphasizing design and innovation in its brand messaging.

The TT's influence extended even to Audi's motorsport programs. The car's popularity led to increased interest in Audi's racing initiatives, particularly in touring car racing, where TT-inspired vehicles competed successfully. This renewed focus on motorsport helped reinforce Audi's performance credentials, creating a virtuous cycle in which racing success fed into the appeal of road cars.

In essence, the first-generation TT did more than just add a successful model to Audi's lineup; it redefined what an Audi could be. It challenged preconceptions about the brand and opened up new possibilities in design and engineering. The TT's impact on Audi's lineup was transformative, elevating the entire brand and setting the stage for Audi's continued success in the luxury car market. The principles established with the TT, innovative design, advanced technology, and a focus on the driving experience, continue to guide Audi's approach to vehicle development to this day.

Section 4.7: Legacy of the First Generation

The first-generation Audi TT left a lasting mark on the automotive world, creating a legacy that extends far beyond its production years. As time has passed, this fantastic sports car has transitioned from a contemporary marvel to a cherished classic, with its influence still resonating in the industry today. In the collector car market, well-maintained first-generation TTs have begun to appreciate, with in-demand editions and low-mileage examples.

Enthusiasts and collectors alike recognize the historical significance of these vehicles, viewing them as pivotal designs that marked a turning point in automotive aesthetics. The TT quattro Sport and early production models are especially sought after, commanding premium prices and often featuring in prestigious classic car auctions.

The TT's influence extended far beyond Audi, inspiring designs from competing manufacturers. In the years following its debut, many sports cars began adopting similar design cues, such as clean lines, minimalist interiors, and bold, simple shapes. The TT's impact can be seen in vehicles ranging from compact sports cars to luxury sedans, proving the universality of its design language.

In popular culture, the TT quickly became a style icon of the turn of the millennium. Its shape made it a favorite in movies, TV shows, and music videos, often used to signify modernity, sophistication, and forward-thinking design. This cultural penetration helped cement the TT's status not just as a car, but as a symbol of an era's aesthetic ideals.

The enduring appeal of the first-generation TT is evident in its continued popularity among enthusiasts. Many consider it the purest expression of the TT concept, appreciating its uncompromising commitment to design principles. Owners' clubs and forums dedicated to the first-generation TT remain active, with members sharing restoration tips, organizing meets, and celebrating the timeless design and engaging driving experience that define these vehicles.

In the realm of automotive design education, the TT has become a case study in successful concept-to-production translation. Design schools often use the TT as an example of how to maintain a concept car's vision while adapting it for real-world production and use. Students analyze everything from its proportions to its material choices, learning valuable lessons about the balance between aesthetics and practicality in automotive design.

The first-generation TT's legacy also lives on in Audi's current lineup. While subsequent generations of the TT evolved the design, they all pay homage to the original's pioneering aesthetic. Moreover, elements of the first TT's design philosophy, such as its emphasis on clean lines, high-quality materials, and innovative technology, continue to influence Audi's broader design language across its entire range.

Perhaps most significantly, the TT demonstrated the power of design to transform a brand's image and market position. Its success gave Audi the confidence to take more risks in design and engineering, leading to a period of innovation that helped elevate the brand to its current premium status.

Looking back on the first-generation Audi TT, it's clear that its impact transcended its role as a sports car. It was a design revolution, a cultural touchstone, and a bold statement of automotive philosophy. The TT proved that a production car could be as visually striking and emotionally engaging as any concept, setting a new standard for the fusion of art and engineering in the automotive world. Its legacy serves as a testament to the enduring power of innovative design, continuing to inspire and influence the industry more than two decades after its introduction.

The Audi TT: A Journey Through Design and Innovation

Chapter 5: Performance and Handling: The TT's Driving Dynamics

Section 5.1: The Heart of the Beast - TT's Powertrains

At the core of every Audi TT lies a beating heart of engineering excellence, its powertrain. Since its inception, the TT has undergone remarkable evolution in its engine offerings, with each generation pushing the boundaries of performance and efficiency.

The journey began with the renowned 1.8T engine in the first-generation TT. This turbocharged four-cylinder powerplant set the stage for future performance, offering a blend of responsiveness and efficiency that would become a hallmark of the TT lineup. Available in various states of tune, from 180 to 225 horsepower, it provided a solid foundation for the TT's sporty character.

As the TT evolved, so did its engine options. The second generation introduced a wider range of powerplants, including a naturally aspirated 3.2-liter V6 for those craving a more traditional sports car sound and power delivery. However, it was the continued refinement of the turbocharged four-cylinder engines that truly defined the TT's character, culminating in the high-output versions found in the TTS models.

The third-generation TT, launched in 2014, introduced a new level of powertrain sophistication. The base model's 2.0-liter TFSI engine now produces 228 horsepower, while the TTS boasts an impressive 288 horsepower from the same displacement. At the top of the range, the TT RS showcases Audi's engineering prowess with its 2.5-liter five-cylinder turbocharged engine, delivering a whopping 394 horsepower and an exhaust note that pays homage to Audi's rallying heritage.

Central to the TT's performance capabilities is Audi's renowned Quattro all-wheel drive system. This sophisticated technology can distribute power between the front and rear axles in milliseconds, ensuring optimal traction in all driving conditions. Whether you're launching from a standstill, cornering at high speeds, or navigating treacherous weather, the Quattro system provides a level of confidence and capability that sets the TT apart from its rear-wheel-drive competitors.

The evolution of the TT's transmission options has been equally impressive. While early models featured traditional manual gearboxes that provided an engaging, hands-on driving experience, later generations introduced Audi's S tronic dual-clutch transmission. This technological marvel offers lightning-fast gear changes, shaving precious seconds off acceleration times while still allowing for manual control when desired. The S tronic transmission has become so quick and efficient that it's now the preferred choice for many performance enthusiasts, offering the best of both worlds - the engagement of a manual and the speed of an automatic.

One of the key factors in the TT's impressive performance is its favorable power-to-weight ratio. Audi's engineers have consistently focused on keeping the TT's weight in check, even as safety requirements and feature content have increased over the years. The use of aluminum in the car's construction has been a breakthrough, particularly in the third generation. The base model of this generation weighs just 2,712 lbs, allowing its engines to deliver thrilling acceleration and nimble handling.

The results of this continuous evolution are evident in the TT's performance figures. The latest TT models showcase impressive acceleration, with the TT RS capable of sprinting from 0-60 mph in just 3.6 seconds - a figure that would have been the domain of supercars not too long ago. Top speeds have also increased, with some variants capable of reaching electronically limited top speeds of 155 mph and even higher when electronically limited.

Throughout its lifetime, the Audi TT's powertrains have been a testament to the brand's commitment to performance and innovation. From the early turbocharged four-cylinders to the latest high-output five-cylinder engines, each generation has raised the bar, delivering more power, better efficiency, and an ever-more thrilling driving experience. Coupled with the Quattro all-wheel drive system and advanced transmission options, the TT's powertrains form the foundation of a true driver's car, ready to excite and engage at every turn of the key.

Section 5.2: Precision Handling - Suspension and Chassis

The Audi TT's reputation as a driver's car is mainly due to its exceptional handling, the result of careful engineering and continuous refinement of its suspension and chassis systems. From the first generation to the latest models, Audi has consistently pushed the boundaries of what's possible in a compact sports car, creating a vehicle that's as comfortable on winding mountain roads as it is on the racetrack.

At the front, the TT has always used a MacPherson strut suspension. This setup offers several advantages, particularly for a sports car like the TT. The MacPherson strut design allows for a lower hood line, contributing to the TT's silhouette and improving aerodynamics. It also provides excellent stability and precise steering control, crucial factors in the TT's responsive handling. The compact nature of this suspension type also allows for more space in the engine bay, facilitating easier maintenance and potential performance upgrades.

The rear suspension of the TT has seen significant evolution over its generations, marking one of the most substantial improvements in the car's handling dynamics. The first-generation TT utilized a torsion-beam rear suspension, which, while adequate for many driving scenarios, had limitations for high-performance handling.

The big leap came with the second generation, which introduced a sophisticated multi-link rear suspension. This switch to a multi-link design dramatically improved the TT's handling characteristics, offering better wheel control, reduced body roll, and enhanced stability during high-speed cornering. The multi-link setup allows each wheel to respond to road imperfections independently, improving both ride comfort and handling precision.

One of the most innovative features in the TT's suspension arsenal is the optional Magnetic Ride system. This adaptive suspension technology uses magnetically charged particles suspended in fluid within the shock absorbers. By applying an electrical charge, the system can instantly alter the fluid's viscosity, thereby changing the suspension's damping rates.

What's truly remarkable is the speed at which this system operates. It can adjust damping rates up to 1,000 times per second. This allows the TT to provide an optimal balance of comfort and performance, adapting to road conditions and driving style in real-time. In comfort mode, the system can absorb road imperfections for a smooth ride, while in dynamic mode, it can stiffen up to minimize body roll and maximize cornering performance.

Chassis stiffness is another area where Audi has made significant strides with each generation of the TT. A stiffer chassis provides numerous benefits, including improved handling precision, better road feedback, and reduced noise, vibration, and harshness (NVH). The third-generation TT, for example, boasts a chassis that is

23% stiffer than its predecessor. This increase in rigidity results in more precise handling, allowing the suspension to work more effectively and providing the driver with better feedback through the steering wheel and seat. The stiffer chassis also contributes to improved safety, better long-term durability, and a more premium feel overall.

Weight distribution and balance play crucial roles in the TT's handling dynamics. Despite being based on a front-wheel-drive platform, Audi has worked hard to achieve a near 50:50 weight distribution in the TT, particularly in Quattro all-wheel-drive models. This is accomplished through careful component placement and the use of lightweight materials in strategic areas. The TT's front-engine, all-wheel-drive layout provides excellent traction and stability. At the same time, the balanced weight distribution enhances cornering ability and reduces the tendency for understeer – a common issue in front-heavy cars.

The use of lightweight materials throughout the TT's construction further enhances its handling capabilities. Aluminum features prominently in the TT's body and chassis components, significantly reducing overall weight without compromising structural integrity. This weight reduction improves all aspects of performance, from acceleration and braking to cornering and fuel efficiency.

In conclusion, the Audi TT's exceptional handling results from a holistic approach to chassis and suspension design. From its MacPherson strut front suspension and advanced multi-link rear setup to its adaptive Magnetic Ride system and lightweight, rigid chassis, every element works in harmony to deliver a driving experience that is both exhilarating and refined.

It's this attention to detail and continuous improvement that has kept the TT at the forefront of handling performance in its class, generation after generation.

Section 5.3: Stopping Power - Braking Systems

The Audi TT's exhilarating performance isn't just about acceleration and handling; it's equally about the ability to bring the car to a halt quickly and confidently. Over the years, Audi has continuously refined the TT's braking systems, ensuring they match the car's increasing power and performance capabilities.

From the first generation to the latest models, the evolution of brake technology in the TT has been remarkable. Early models featured solid rotors, which were soon replaced by ventilated discs for improved heat dissipation. As the TT's performance increased, so did the size and sophistication of its brakes. The TT RS, for instance, boasts massive 14.6-inch front rotors paired with 8-piston calipers, providing phenomenal stopping power that can bring the car from high speeds to a standstill in impressively short distances.

But it's not just about raw stopping power. The TT's braking system is a masterpiece of engineering, incorporating advanced technologies to enhance both safety and performance. The Anti-lock Braking System (ABS) and Electronic Brake Force Distribution (EBD) work in tandem to optimize braking performance in various conditions. The TT's advanced ABS can adjust brake pressure individually to each wheel, optimizing stopping distances on multiple surfaces. This is particularly useful when driving on mixed road conditions or during emergency maneuvers.

One of the most innovative features in the TT's braking system is the brake-based torque vectoring. This clever system enhances cornering performance by applying slight brake pressure to the inside wheels during hard cornering.

The result is a car that rotates more effectively through turns, offering a more agile, responsive driving experience. It's a perfect example of how Audi uses braking technology not just for stopping, but as an active part of the car's handling dynamics.

Audi's engineers have paid particular attention to brake feel and pedal feedback, recognizing that these elements are crucial to driver confidence, especially in high-performance driving. The brake pedal in the TT is tuned to provide progressive resistance, allowing for precise modulation during performance driving. This level of control is essential when pushing the car to its limits on a track or during spirited driving on winding roads.

For those who demand even more from their TT, especially for track use, Audi offers track-focused braking options. The TT RS, for example, is available with optional carbon-ceramic brakes. These provide fade-free performance even under extreme track conditions, where repeated heavy braking can overwhelm conventional steel brakes. While they come at a premium, carbon-ceramic brakes offer unparalleled performance and longevity for the most demanding drivers.

It's worth noting that Audi's focus on braking performance extends beyond just the hardware. The integration of the braking system with the car's electronic stability control and traction control systems creates a holistic approach to vehicle dynamics. This integration ensures that the TT's braking performance complements its acceleration and handling characteristics, resulting in a car that feels balanced and predictable at the limits of adhesion.

In conclusion, the Audi TT's braking systems are a testament to the car's performance pedigree. From the substantial stopping power of its large brake rotors to the sophisticated electronic systems that optimize brake force distribution, every aspect has been carefully engineered to provide confidence-inspiring performance.

Whether it's providing fade-free braking on a track day or ensuring safe stops in everyday driving, the TT's brakes are an integral part of its high-performance character, truly embodying the phrase "stopping power."

Section 5.4: Steering Precision - From Hydraulic to Electric

The evolution of the Audi TT's steering system is a testament to the relentless pursuit of perfection in automotive engineering. As we delve into this crucial aspect of the TT's performance, we'll explore how Audi has maintained the car's reputation for precise handling while embracing new technologies.

The transition from hydraulic to electric power steering marks a significant milestone in the TT's history. While purists initially resisted this change, the switch to electric power steering allowed for more precise tuning and better fuel efficiency. Audi's engineers worked tirelessly to ensure that the electric system could replicate and even enhance the feel and responsiveness that drivers had come to expect from the TT.

One of the key advantages of the electric power steering system is its ability to offer a variable-ratio steering rack. This innovative feature quickens the steering ratio as you turn the wheel further, providing agility in tight corners without sacrificing stability at higher speeds. The result is a car that feels nimble and responsive in urban environments while maintaining rock-solid stability on the autobahn.

Despite being electric, the TT's steering provides ample feedback, keeping the driver connected to the road surface. This was a crucial consideration for Audi's engineers, who understood that the TT's appeal lies in its ability to engage the driver. They achieved this by carefully calibrating the electric motor's assistance and implementing sophisticated software algorithms that can simulate road feel.

The integration of the electric power steering system with modern driver assistance features is another area where the TT shines. Features like lane keeping assist can now be implemented without compromising the TT's sporty character. This integration showcases Audi's ability to balance performance with safety and convenience.

For those seeking the ultimate track experience, Audi offers track-focused steering calibrations in its high-performance variants. The TT RS, for example, features a more aggressive steering calibration, providing quicker turn-in for track driving. This allows enthusiasts to exploit the car's full potential on the circuit while still maintaining a comfortable and manageable ride for daily use.

The steering wheel itself has also evolved, with Audi paying close attention to its design and functionality. The flat-bottomed steering wheel, now a signature element of the TT, not only looks sporty but also provides additional thigh clearance, enhancing comfort during spirited driving. The wheel's thickness and grip points have been optimized to provide the best possible control and feedback.

Audi's commitment to steering precision is best exemplified by their extensive testing procedures. Each TT undergoes rigorous evaluation on various road types and conditions, ensuring that the steering feel is consistent and appropriate across all driving scenarios. From the smooth tarmac of a race track to the rough surfaces of rural roads, the TT's steering is designed to inspire confidence.

As we look to the future, Audi continues to refine the TT's steering system. Research into steer-by-wire technology and even more advanced electric systems promises to enhance the car's capabilities further. However, Audi remains committed to preserving the direct and engaging steering feel that has become a hallmark of the TT driving experience.

In conclusion, the evolution of the Audi TT's steering system from hydraulic to electric represents a successful marriage of traditional sports car values with cutting-edge technology. By maintaining a focus on driver engagement while embracing modern engineering, Audi has ensured the TT remains a benchmark for steering precision in its class.

Section 5.5: Tires and Wheels - Where Rubber Meets Road

The Audi TT's performance isn't just about what's under the hood; it's also about where the rubber meets the road. The tires and wheels play a crucial role in the car's handling, grip, and overall driving experience. Over the years, Audi has continually refined these components to enhance the TT's performance and aesthetic appeal.

The evolution of tire technology has been a key factor in the TT's improved performance across generations. In the latest models, Audi has partnered with top tire manufacturers to develop specially formulated compounds that offer an optimal balance of grip, longevity, and ride comfort. For example, the latest TT models feature specially developed Hankook tires, optimized for both performance and comfort. These tires provide excellent traction in various conditions while also contributing to the car's refined road manners.

Wheel design in the TT lineup has always been a perfect blend of form and function. The five-spoke wheel design, a hallmark of many Audi models, is more than just a visual treat. This design also serves a practical purpose by aiding in brake cooling. The open-spoke pattern allows air to flow freely to the brake rotors, helping dissipate heat during spirited driving or track sessions. This attention to detail exemplifies Audi's commitment to performance, even in seemingly aesthetic choices.

When it comes to tire options, Audi offers both summer and all-season variants for the TT, each with its own set of advantages. Summer tires provide ultimate grip in warm conditions, allowing drivers to exploit the car's performance potential fully. These tires feature softer compounds and more aggressive tread patterns, resulting in superior cornering abilities and shorter braking distances in dry conditions. However, their performance significantly drops in colder temperatures.

On the other hand, all-season tires extend the TT's usability year-round. While they may not match the outright performance of summer tires in ideal conditions, they offer a more versatile solution for drivers who face a variety of weather conditions. These tires provide a good balance of performance, comfort, and longevity, making them a popular choice for daily drivers.

The wheel size also plays a significant role in the TT's handling characteristics. Audi offers a range of wheel sizes, typically from 17 inches up to 20 inches, depending on the model and trim level. Larger wheels, such as the 20-inch options available on the TT RS, provide a more aggressive stance and sharper turn-in response. The increased wheel diameter allows for larger brake rotors, enhancing stopping power. However, this comes at the cost of some ride comfort, as there's less tire sidewall to absorb road imperfections.

In contrast, smaller wheel sizes offer a more compliant ride and can be beneficial in areas with poor road conditions. They also generally provide better performance in winter conditions when fitted with appropriate tires. The choice of wheel size often comes down to a balance between performance desires and practical considerations.

Audi has also addressed tire emergencies in the TT. Some models offer run-flat tires, a technology that allows drivers to continue for up to 50 miles after a puncture. This feature provides peace of mind, especially considering that many modern cars, including some TT models, have done away with spare tires to save weight and increase cargo space. For those TT models that don't come with run-flat tires, Audi provides a space-saver spare or a tire repair kit. While not ideal for long-distance use, these solutions offer a temporary fix to get drivers to a service center safely.

The choice of tires and wheels on the Audi TT goes far beyond mere aesthetics. It's a crucial aspect of the car's performance envelope, affecting everything from acceleration and braking to cornering and ride comfort. By carefully selecting and continually improving these components, Audi ensures that the TT delivers on its promise of sporty performance without compromising daily usability. Whether you're carving through mountain roads on a summer day or commuting through a rainy city, the TT's tires and wheels work in harmony with its other systems to provide a driving experience that's both exhilarating and confident.

Section 5.6: Electronic Aids - Enhancing Performance and Safety

The Audi TT's evolution has been marked not only by mechanical improvements but also by the integration of sophisticated electronic aids that enhance both performance and safety. These systems have become increasingly advanced, allowing drivers to push the limits of the car's capabilities while maintaining a crucial safety net.

At the heart of the TT's electronic assistance is its traction and stability control system. Over the years, this technology has become remarkably sophisticated. Modern TTs feature multi-stage stability control, allowing drivers to tailor intervention levels to their skill and the driving conditions. This customization ensures the car can cater to both novice drivers seeking maximum safety and experienced enthusiasts seeking more dynamic handling. In its most permissive settings, the system allows for a degree of slip and rotation that can be exhilarating on a twisty road or track, while still providing a safety backup if things get too wild.

One of the most impressive electronic features in high-performance TT models is the launch control system. This technology allows drivers to achieve optimal acceleration from a standing start, consistently and reliably. With launch control activated, even the most powerful TT RS can consistently achieve its claimed 0-60 time, even in challenging conditions. The system manages engine rpm, clutch engagement (in dual-clutch models), and power distribution to ensure maximum traction and acceleration. It's a perfect example of how electronic aids can make the car's performance more accessible to drivers of all skill levels.

The Audi Drive Select system is another electronic feature that significantly affects the TT's performance. This system allows drivers to customize various aspects of the car's behavior, effectively changing its personality at the touch of a button. In Dynamic mode,

The Audi TT: A Journey Through Design and Innovation

for instance, throttle response sharpens, the suspension stiffens, and the exhaust opens up for a more engaging driving experience. Conversely, Comfort mode softens the ride and quiets the exhaust for more relaxed cruising. This adaptability makes the TT a true all-rounder, capable of being both a comfortable daily driver and an exhilarating sports car.

Audi has also leveraged its virtual cockpit technology to enhance the driving experience. The fully digital instrument cluster can display a wealth of performance data, keeping the driver informed during spirited driving. G-force meters, lap timers, and even tire pressure information are all available at a glance. This integration of performance data into the driver's primary field of view allows for quick decision-making without taking eyes off the road – a crucial safety feature during high-performance driving.

Looking to the future, Audi is exploring new ways to use technology to enhance driver performance without diminishing the pure driving experience. There's potential for AI-driven systems that could help drivers improve their skills on track, offering real-time coaching and feedback. However, Audi is careful to balance these aids with the need for driver engagement, ensuring that the TT remains a car for driving enthusiasts.

The evolution of these electronic aids in the TT reflects a broader trend in the automotive industry: the use of technology to make cars safer and more capable while still preserving the thrill of driving. In the TT, these systems work seamlessly in the background, enhancing the car's performance and safety without overshadowing the fundamental joy of driving a well-engineered sports car. They represent a perfect marriage of traditional sports car values with cutting-edge automotive technology, further cementing the TT's position as a thoroughly modern driver's car.

Section 5.7: The TT on Track - From Street to Circuit

The Audi TT has always been more than just a stylish road car. Its performance capabilities shine brightest when pushed to the limit on a race track. This section explores how Audi has honed the TT's track prowess over the years, creating versions that are equally at home on the track as on public roads.

Audi's commitment to track performance is evident in their track-focused variants of the TT. These models feature specific enhancements that elevate the car's capabilities in high-performance driving. For instance, the TT RS with the Dynamic Plus package is a prime example of Audi's track-oriented engineering. This variant includes a carbon fiber engine cover, which not only reduces weight but also adds a touch of racing flair when you pop the hood. The package also features OLED taillights, providing a look and improved visibility. Perhaps most importantly for track enthusiasts, it raises the electronically limited top speed from 155 mph to an impressive 174 mph, allowing drivers to exploit the car's potential on long straights fully.

One of the key challenges in translating street performance to the track is managing heat. Audi has addressed this with sophisticated cooling systems and heat management solutions. Additional radiators and oil coolers in performance models help maintain optimal temperatures even under the extreme conditions encountered during extended track sessions. This attention to thermal management ensures that the TT can deliver consistent performance lap after lap, without the power-sapping heat soak that plagues less track-focused vehicles.

Aerodynamics play a crucial role in high-speed stability, an essential factor for track driving. Audi has continuously refined the TT's aerodynamic profile to enhance its performance at speed. A standout feature is the retractable rear spoiler found on modern TTs. This clever piece of engineering deploys automatically at high speeds, increasing rear downforce by up to 50 kg. This additional downforce improves stability in high-speed corners and under heavy braking, giving drivers the confidence to push harder on the track.

Suspension tuning is another area where Audi has paid special attention to the TT's track capabilities. While the standard TT offers a commendable balance of comfort and performance for road use, track-focused models often feature more aggressive setups. These can include stiffer springs and bushings, as well as more aggressive alignment settings. The result is improved cornering ability, with sharper turn-in, reduced body roll, and enhanced overall grip. These track-tuned suspensions allow drivers to carry more speed through corners and provide better feedback, crucial for finding those last few tenths of a second on a lap time.

Audi's commitment to the TT's performance credentials extends beyond road-legal vehicles and into motorsports. The Audi Sport TT Cup, which ran from 2015 to 2017, was a testament to the TT's racing potential. This spec series featured specially prepared TT race cars, allowing drivers to compete on a level playing field and showcasing the model's inherent performance capabilities. The series served as a proving ground for both the car and up-and-coming drivers, with races held as support events for the prestigious Deutsche Tourenwagen Masters (DTM) series.

The experience gained from the TT Cup and other racing programs has fed back into the development of road-going TTs, creating a virtuous cycle of performance improvement. Lessons learned on the track about everything from brake cooling to weight distribution have influenced the design and engineering of subsequent TT models, ensuring that even the standard road cars benefit from Audi's racing expertise.

For the truly dedicated track enthusiast, Audi offers various performance upgrades that can be retrofitted to enhance the TT's track capabilities further. These can range from uprated brake pads and fluids for improved stopping power to more comprehensive packages including adjustable coilover suspensions and engine tuning options. This aftermarket support, combined with the TT's inherent performance, makes it a popular choice for track-day enthusiasts and amateur racers alike.

The Audi TT: A Journey Through Design and Innovation

The Audi TT's journey from street to circuit is a testament to its versatile performance envelope. Whether it's a standard model being driven enthusiastically at a track day or a fully-prepared race car competing in a professional series, the TT has proven its mettle time and again. It's this blend of everyday usability and genuine track capability that cements the TT's status as a true driver's car, capable of delivering thrills whether you're commuting to work or chasing lap times at your local circuit.

Chapter 6: The TT in Motorsports: Racing Heritage and Achievements

Section 6.1: Early Days: The TT's Debut in Motorsports

The Audi TT's journey from stylish road car to competitive track machine began with a bold decision that would shape its future and cement its place in motorsport history. In the late 1990s, as the TT was making waves in the automotive world with its design, Audi's motorsport division saw an opportunity to showcase the car's potential beyond the streets.

The decision to race the TT was not taken lightly. Audi had a rich racing heritage, but the TT was primarily known for its avant-garde styling rather than its performance credentials. However, the company's engineers and motorsport experts recognized the inherent potential in the TT's compact size, balanced chassis, and quattro all-wheel-drive system. They believed that with the proper modifications, the TT could be transformed into a formidable racing machine.

Preparations for the TT's racing debut were intensive and meticulous. The road-going version underwent significant modifications to meet racing regulations and to enhance its performance capabilities. The interior was stripped down to reduce weight, and a roll cage was installed to ensure driver safety. The suspension was overhauled with stiffer springs and adjustable dampers to improve handling on the track. The engine, already potent in road-going form, was tuned to produce significantly more power, often pushing beyond 300 horsepower in race trim.

Aerodynamics played a crucial role in the TT's transformation. Engineers developed a comprehensive aero package, including a large rear wing, front splitter, and side skirts, all designed to increase downforce and improve stability at high speeds. These modifications not only enhanced the TT's performance but also gave it a more aggressive appearance that hinted at its racing aspirations.

The TT made its first competitive appearances in various national touring car championships and endurance races. Its debut was met with a mix of curiosity and skepticism from the racing community. Many wondered how a car known primarily for its looks would fare against purpose-built racing machines. However, the TT quickly silenced doubters with its impressive performance on the track.

In its early races, the TT demonstrated surprising agility and speed. Its compact size allowed it to navigate tight corners with ease, while its quattro system provided excellent traction, particularly in wet conditions. The car's unique silhouette made it instantly recognizable on the track, and its performances began to draw attention from both fans and competitors alike.

The initial reception from the racing community was one of growing respect. As the TT consistently performed well and even secured some podium finishes, it became clear that Audi had successfully transformed their style icon into a genuine racing contender. Competitors began to take notice, and soon, the TT was being regarded as a serious threat in its class.

These early races provided invaluable lessons for Audi's motorsport division. The team quickly learned about the TT's strengths and weaknesses in a competitive environment. They discovered that while the car excelled in handling and traction, there was room for improvement in straight-line speed and brake cooling. These insights led to continuous refinements and updates to the racing TT, a process that would continue throughout its motorsport career.

The lessons learned from these early races had a profound impact on the development of future TT models, both on the track and for the road. The knowledge gained about aerodynamics, weight distribution, and power delivery would influence the design and engineering of subsequent generations of the TT, ensuring that the racing DNA became an integral part of the car's identity.

As the TT established itself in motorsport, it began attracting talented drivers who recognized its potential. These drivers, with their skill and feedback, played a crucial role in further developing the TT's racing capabilities. Their experiences behind the wheel helped Audi refine the car, making it more competitive with each passing race.

The early days of the TT in motorsports laid a solid foundation for a successful racing career. From its first appearance on the track, the TT proved that it was more than just a pretty face. It demonstrated that with the right engineering and determination, a road car could be transformed into a formidable racing machine. This period marked the beginning of a new chapter in TT's story, one that would see it evolve from a design icon to a respected competitor in motorsports.

Section 6.2: The TT in Touring Car Championships

The Audi TT's foray into touring car championships marked a significant milestone in its motorsport journey. As the TT gained traction on the track, it began to make waves in various national touring car series, showcasing its potential as a formidable competitor.

Audi's decision to enter the TT in touring car championships was a strategic move to prove the car's performance capabilities and enhance its sporty image. The TT's compact size and agile handling made it well-suited for the tight turns and close-quarters racing typical of touring car events.

One of the TT's most notable performances came in the German VLN Endurance Championship, where it competed against a field of seasoned touring cars. The TT's nimble handling and responsive steering allowed it to navigate the challenging Nürburgring Nordschleife with impressive speed and precision. In its class, the TT consistently finished in top positions, often punching above its weight against more established competitors.

The British Touring Car Championship (BTCC) also saw the TT make its mark. Audi Sport UK fielded a modified TT that quickly became a fan favorite due to its shape and impressive on-track performance. The TT's participation in the BTCC not only garnered points and podium finishes but also significantly raised its profile among British motorsport enthusiasts.

As the TT competed in these high-profile series, it underwent continuous technical development. Engineers worked tirelessly to optimize the car's aerodynamics, often incorporating lessons learned from each race weekend. The touring car version of the TT featured a significantly lowered suspension, wider track, and an aggressive body kit that improved downforce and stability at high speeds.

One of the most significant technical advancements driven by touring car racing was the refinement of the TT's quattro all-wheel-drive system. The demands of touring car racing pushed Audi's engineers to develop a more responsive and efficient AWD system, capable of delivering power precisely where it was needed most during high-speed cornering and acceleration out of turns.

When compared to its competitors in the same class, the TT often stood out for its unique combination of all-wheel-drive traction and compact size. While it may have lacked the raw power of some larger-engined rivals, the TT's balanced chassis and excellent corner exit speeds made it a consistent threat on twisty circuits.

The impact of touring car racing on the TT's road-going versions was substantial. Many of the aerodynamic improvements and suspension tweaks developed for the track found their way into production models. The TT RS, in particular, benefited greatly from the motorsport program, with its high-performance five-cylinder engine and advanced AWD system drawing direct inspiration from the race cars.

Moreover, success in touring car championships significantly boosted the TT's credibility as an actual sports car. What was once seen primarily as a style-oriented coupe was now recognized as a capable performance machine. This shift in perception led to increased interest from enthusiasts and helped justify the development of even more potent road-going versions.

The TT's participation in touring car championships also had a profound effect on Audi's overall motorsport strategy. The success and lessons learned from the TT program paved the way for Audi to expand its presence in other touring car series with different models, further cementing the brand's reputation as a builder of fast, reliable, and competitive sports cars.

As the TT continued to evolve both on and off the track, its touring car exploits became an integral part of its story. The championships it competed in served as a proving ground for technology, a showcase for the car's capabilities, and a source of pride for Audi enthusiasts worldwide. The TT's journey through the world of touring car racing demonstrated that this sporty design was backed by genuine performance, forever changing its image from merely a stylish coupe to an actual track-worthy sports car.

Section 6.3: Endurance Racing: Testing the TT's Limits

The world of endurance racing presented a unique challenge for the Audi TT, pushing its performance and reliability far beyond what was required in shorter race formats. As Audi sought to prove the TT's mettle in the grueling world of long-distance motorsports, the car underwent significant modifications to meet the demands of races that could last 6 to 24 hours.

To compete in endurance events, the TT required extensive alterations to its powertrain, suspension, and aerodynamics. Engineers focused on enhancing fuel efficiency without sacrificing power, a crucial factor in minimizing pit stops during lengthy races. The engine was often detuned slightly from its sprint racing configuration to prioritize longevity and consistent performance over outright speed. Cooling systems were upgraded to handle the prolonged stress, and larger fuel tanks were installed to extend the car's range between refueling stops.

The TT's participation in endurance racing yielded some impressive results, showcasing the car's inherent strengths and the effectiveness of Audi's modifications. One of the most notable performances came at the Nürburgring 24 Hours, where a specially prepared TT RS finished first in its class and achieved a respectable overall position against purpose-built endurance prototypes. This achievement not only demonstrated the TT's capability but also its versatility across different racing disciplines.

The rigors of endurance racing had a profound impact on the TT's development, particularly in improving reliability. The punishing nature of these events exposed weaknesses that might not have surfaced in shorter races or on the road. As a result, Audi's engineers identified and addressed potential issues, leading to enhanced durability across both racing and production models. Components such as transmissions, brake systems, and electrical systems all benefited from the lessons learned on the endurance racing circuit.

Driver experiences and testimonials from endurance events provided valuable insights into the TT's performance under extreme conditions. Many drivers praised the car's balanced handling and predictable behavior over long stints, attributes that became increasingly important as fatigue set in during the later hours of a race. The TT's relatively compact size and efficient aerodynamics also proved advantageous on twisting circuits, allowing it to navigate through traffic with agility.

One TT endurance driver, after completing a grueling 12-hour race, remarked, "The car felt as responsive and composed in the final laps as it did at the start. It's a testament to the TT's engineering that it can maintain such consistency under these punishing conditions."

Data gathered from endurance racing played a crucial role in refining the TT's systems to improve longevity and performance. Transmission cooling was enhanced, leading to more robust gearboxes in production models. Brake fade, a common issue in endurance racing, was addressed with new materials and cooling designs that eventually made their way into high-performance road versions of the TT.

Audi's commitment to endurance racing with the TT not only improved the car's capabilities but also bolstered its reputation among enthusiasts and potential customers. The sight of a TT battling through the night at Le Mans or conquering the Green Hell of the Nürburgring served as a powerful marketing tool, showcasing the model's performance credentials in the most demanding conditions imaginable.

As the TT continued to prove itself in endurance events, it gained respect within the racing community. It helped dispel any lingering notions that it was merely a stylish boulevard cruiser. The lessons learned and successes achieved in these grueling races contributed significantly to the TT's evolution, ensuring that every generation of the road car benefited from the extremes of endurance motorsport.

Section 6.4: The TT in Hill Climb and Time Attack Events

As the Audi TT continued to prove its mettle across various motorsport disciplines, it found a particularly suitable niche in hillclimb competitions and time attack events. These challenging formats allowed the TT to showcase its agility, power, and precision handling in ways that traditional circuit racing couldn't always highlight.

Adapting the TT for hill climb competitions required a unique approach. Engineers focused on maximizing power-to-weight ratio and enhancing the car's responsiveness for the tight, twisting ascents typical of hill climb courses. Suspension systems were fine-tuned to provide optimal grip on varying road surfaces, while aerodynamic modifications helped keep the car planted at high speeds on steep inclines. The quattro all-wheel-drive system, a hallmark of Audi performance, proved to be a significant advantage in these events, providing superior traction and stability.

The TT's hill climb performances quickly garnered attention in the motorsport world. At events like the famous Pikes Peak International Hill Climb, specially prepared TTs demonstrated remarkable speed and control. One particularly memorable run saw a modified TT RS complete the treacherous 12.42-mile course in under 11 minutes, a time that put it in competition with purpose-built race cars. This achievement not only showcased the TT's capabilities but also solidified its reputation as a serious performance machine.

In the realm of time attack events, where cars compete to set the fastest lap time on a closed circuit, the TT found another arena to shine. These events allowed for more extreme modifications than many traditional racing series, and Audi's engineers took full advantage. Time attack TTs featured dramatically enhanced aerodynamics, including massive rear wings and front splitters, along with significantly boosted engine outputs often exceeding 600 horsepower.

The technical innovations developed for these high-intensity, short-duration events pushed the boundaries of what was possible with the TT platform. Advanced engine management systems, cutting-edge turbocharger technology, and exotic materials all found their way into these specialized machines. While many of these modifications were too extreme for road use, they served as valuable test beds for technologies that would eventually trickle down to production models.

The most significant impact of the TT's participation in hill climb and time attack events was on its performance image. These disciplines, with their emphasis on outright speed and handling precision, aligned perfectly with the TT's design philosophy. Success in these events helped transform the perception of the TT from a stylish sports coupe into a legitimate performance car capable of competing with the best in the world.

The sight of TTs conquering mountain roads or setting blistering lap times became a powerful marketing tool for Audi. It demonstrated that the car's performance capabilities extended far beyond its elegant exterior, appealing to a new demographic of hardcore driving enthusiasts. This racing pedigree influenced the development of subsequent TT models, with each generation becoming more track-focused and performance-oriented.

Enthusiasts and the automotive press took notice of the TT's prowess in these events. Magazine features and online videos showcasing the car's hill climb and time attack exploits helped build a cult following. This grassroots support translated into increased interest in both the racing program and the road-going TT models.

The experience gained from hillclimb and time attack competitions had a lasting impact on the TT's development. Lessons learned about weight distribution, aerodynamics, and power delivery informed the design of future models. Even for everyday drivers who never set foot on a racetrack, the benefits of this racing heritage could be felt in the car's responsive handling and confident performance.

As the TT continued to evolve, its success in hill climb and time attack events became an integral part of its story. These disciplines allowed the car to demonstrate its versatility and performance potential, contributing significantly to its reputation as a true driver's car. From winding mountain roads to purpose-built circuits, the TT proved time and again that it was more than capable of punching above its weight, earning respect and admiration from motorsport enthusiasts around the world.

The Audi TT: A Journey Through Design and Innovation

Section 6.5: Racing Success Stories and Memorable Moments

The Audi TT's journey in motorsports is punctuated by a series of triumphant victories and unforgettable moments that have secured its place in racing history. These success stories not only showcase the TT's capabilities on the track but also highlight the skill and determination of the drivers who piloted these machines to glory.

One of the most significant achievements in the TT's racing career came in 2002 when Frank Biela piloted a modified TT to victory in the German Touring Car Championship (DTM). This win was particularly notable as it marked the first time a compact sports car had triumphed in this prestigious series, traditionally dominated by larger sedans. Biela's victory showcased the TT's agility and power, proving that it could compete with and surpass more established racing platforms.

Another standout driver who made his mark with the TT was Pierre Kaffer. In the 2004 ADAC VLN Endurance Racing Championship at the Nürburgring, Kaffer demonstrated the TT's endurance capabilities by securing multiple class wins. His consistent performances throughout the season highlighted the TT's reliability and competitiveness in long-distance racing formats.

The TT's racing success wasn't limited to European circuits. In North America, the car made waves in the SPEED World Challenge GT series. Randy Pobst, a renowned sports car racer, piloted a TT to several podium finishes in the mid-2000s, including a memorable victory at Mosport International Raceway in 2005. Pobst's skillful driving, combined with the TT's nimble handling, allowed him to outmaneuver larger, more powerful competitors on the twisting Canadian circuit.

One of the most dramatic moments in TT racing history occurred during the 2006 24 Hours of Nürburgring. The Raeder Motorsport team's TT, driven by Marc Hennerici, Frank Schmickler, and Philip Leisen, was locked in a fierce battle for the SP3T class lead. With just hours to go, a sudden rainstorm hit the track. The TT's quattro all-wheel-drive system proved invaluable, allowing the team to maintain control in treacherous conditions while their competitors struggled. This advantage led to a come-from-behind victory that showcased the TT's all-weather performance capabilities.

The TT has also set impressive records in various racing disciplines. In 2015, a heavily modified TT RS, driven by Hans-Joachim Stuck, set a new lap record for compact cars at the Hockenheimring during a Sport Auto event. The blistering time of 1:09.040 demonstrated the TT's potential when pushed to its absolute limits, surpassing expectations for a car of its size and class.

Media coverage of the TT's racing successes has been consistently positive, often highlighting the car's ability to punch above its weight class. Automotive publications and racing periodicals have praised the TT's versatility and competitiveness across various racing formats. This media attention has played a crucial role in shaping public perception of the TT, transforming its image from that of a stylish road car to a serious performance machine.

The public reaction to the TT's racing achievements has been one of admiration and increased respect. Enthusiasts who may have initially dismissed the TT as a fashion-focused sports car were forced to reconsider their stance in light of its on-track prowess. This shift in perception has not only boosted the TT's credibility among hardcore automotive fans but has also attracted a new demographic of buyers who appreciate both style and substance.

These racing success stories and memorable moments have contributed significantly to the TT's legacy. They serve as testaments to the car's engineering excellence and Audi's commitment to performance. More importantly, they have inspired generations of TT owners and enthusiasts, fostering a deep connection between the brand and its supporters. The TT's racing achievements continue to be celebrated at enthusiast meetups, in online forums, and through various media channels, ensuring that the car's competitive spirit lives on long after the checkered flag has fallen.

Section 6.6: From Track to Street: Racing-Inspired Road Models

The Audi TT's racing success didn't stay on the track; it found its way onto public roads through a series of exciting racing-inspired special editions. Audi's engineers and designers worked tirelessly to bring the thrill of motorsport to everyday drivers, creating a tangible link between the TT's racing pedigree and its street-legal counterparts.

One of the most notable outcomes of this track-to-street transfer was the introduction of racing-inspired special editions. These limited-run models captured the essence of the TT's competitive spirit, featuring unique styling cues, performance enhancements, and exclusive badging. For instance, the TT RS GT, unveiled in 2013, was a prime example of how Audi translated racing success into a road-going masterpiece. Limited to just 99 units worldwide, this model boasted a more powerful engine, a stripped-down interior reminiscent of a race car, and aerodynamic improvements derived directly from the track.

The technology transfer from race cars to production models was equally impressive. Audi's engineers meticulously analyzed data and lessons from countless races, using this valuable information to improve the road-going TT. Advancements in areas such as engine management, suspension tuning, and aerodynamics were carefully adapted for street use, ensuring that even the standard TT models benefited from the rigors of competition.

Performance enhancements derived from racing experience became a hallmark of the TT lineup. The quattro all-wheel-drive system, already a staple of the TT, saw continuous refinement based on its performance in various racing conditions. Improved traction control systems, more responsive steering, and enhanced brake systems were all byproducts of the TT's time on the track. These upgrades not only improved the car's performance but also enhanced its safety and handling characteristics for everyday drivers.

The Audi TT: A Journey Through Design and Innovation

Audi's marketing department recognized the immense value of the TT's racing pedigree and leveraged it effectively. Advertisements and promotional materials often featured the TT in racing livery or showcased its track achievements. This strategy helped to position the TT not just as a stylish sports car, but as a serious performance machine with proven capabilities. Special events and driving experiences were organized to allow customers to experience the TT's racing DNA firsthand, further cementing its reputation as a driver's car.

The customer reception of these race-inspired road models was overwhelmingly positive. Enthusiasts and casual buyers alike appreciated the connection to motorsport, seeing it as a badge of honor that set the TT apart from its competitors. The racing-inspired models often sold out quickly, becoming instant collectibles. More importantly, the halo effect of these special editions boosted the appeal of the entire TT range, with customers drawn to the idea of owning a piece of Audi's racing legacy.

The success of these racing-inspired road models did more than just boost sales; it created a feedback loop that continued to influence the development of both road and race versions of the TT. As customers demanded greater performance and race-derived features, Audi pushed the boundaries of what was possible in a road-legal sports car. This, in turn, provided valuable insights that could be applied back to the racing program, creating a virtuous cycle of innovation and improvement.

In essence, the racing-inspired road models of the Audi TT became the ultimate expression of the car's dual nature - a stylish road car with the heart of a racer. They served as a testament to Audi's commitment to performance and innovation, bridging the gap between motorsport and the daily commute. For TT owners, every drive became an opportunity to experience a little bit of racing glory, turning ordinary roads into personal racetracks.

Section 6.7: The Legacy of TT in Motorsports

The Audi TT's venture into motorsports has left an indelible mark on both the model's history and Audi's racing legacy. Since its debut on the track, the TT has transformed from a stylish road car into a respected competitor, reshaping its brand image and influencing future Audi performance models.

The impact on the TT's brand image cannot be overstated. What began as a design-focused sports coupe has evolved into a symbol of performance and engineering prowess. The TT's success across various racing disciplines has elevated its status beyond a mere fashion statement, earning it respect among driving enthusiasts and motorsport aficionados alike. This racing pedigree has added depth and credibility to the TT's performance claims, making it a more appealing option for buyers seeking a car with genuine sporting credentials.

The TT's motorsport experience has had a profound influence on future Audi performance models. Lessons learned from pushing the TT to its limits on the track have been applied across Audi's performance lineup, informing everything from aerodynamics to powertrain development. The success of racing-inspired TT variants has also paved the way for more track-focused models across Audi's range, demonstrating the market appetite for road cars with genuine racing DNA.

When compared to other sports cars' racing heritage, the TT holds its own admirably. While it may not have the decades-long racing history of some competitors, the TT has quickly established itself as a formidable presence in touring car championships, endurance racing, and hill climb events. Its success across multiple disciplines showcases the versatility and robust engineering that underpin the TT's design.

The TT's racing exploits have fostered a vibrant enthusiast community. From amateur racers to ardent fans, a passionate group has formed around TT Motorsports. This community shares knowledge, organizes events, and keeps the spirit of TT racing alive, even as older models transition out of professional competition. The enthusiasm of this community has helped maintain the TT's performance image and has likely contributed to its longevity in Audi's lineup.

In the broader context of Audi's racing history, the TT occupies a unique place. While it may not have the headline-grabbing Le Mans victories of Audi's prototype racers, the TT has been crucial in demonstrating the performance potential of Audi's production-based models. It has served as a bridge between Audi's high-end motorsport programs and its road cars, making racing technology and excitement accessible to a broader audience.

The legacy of the TT in motorsports extends beyond trophies and lap times. It represents Audi's commitment to infusing its road cars with racing DNA, proving that performance and practicality can coexist. The TT's racing success has challenged preconceptions about what a sports car can be, showcasing that innovative design and motorsport prowess are not mutually exclusive.

As we reflect on the TT's motorsport legacy, it's clear that its impact reaches far beyond the racetrack. The TT's journey from a design icon to a respected racing competitor has enriched its history, enhanced its appeal, and solidified its place in the pantheon of great sports cars. This racing heritage continues to influence each new generation of the TT, ensuring that every model carries with it the spirit of competition and the thrill of the track.

The Audi TT: A Journey Through Design and Innovation

Chapter 7: The Second Generation: Refining a Classic

Section 7.1: Design Evolution

The second generation of the Audi TT faced the formidable challenge of evolving an iconic design while maintaining its character. Audi's designers approached this task with a delicate balance of reverence for the original and a forward-thinking vision.

Preserving the TT's DNA was paramount in the design process. The wheel arches and sloping roofline, hallmarks of the first generation, remained integral to the new model. These elements ensured the TT's silhouette retained its unmistakable profile, making it recognizable at a glance. Preserving these core design features demonstrated Audi's commitment to the TT's heritage while setting the stage for subtle yet impactful changes.

Sharpening the silhouette became a key focus for the design team. The new TT adopted a more angular and aggressive approach, aligning with Audi's evolving design language.

The front grille, in particular, underwent a significant transformation by adopting Audi's signature single-frame design. This change gave the TT a more assertive face, enhancing its sporty character without sacrificing its elegant proportions. The sharper lines extended throughout the body, from the sculpted hood to the more defined shoulder line, creating a sense of motion even when the car was stationary.

Aerodynamic improvements played a crucial role in the design evolution. The team at Audi understood that form must follow function, especially in a sports car. A retractable rear spoiler was ingeniously integrated into the body, deploying at high speeds to improve downforce without compromising the clean lines when retracted. This feature exemplified how the designers successfully married aesthetics with performance, a hallmark of the TT's philosophy.

Interior refinements further elevated the TT's appeal. The cockpit, always a strongpoint of the TT, received careful attention. The baseball-stitched leather seats, a beloved feature of the first generation, were refined for enhanced comfort and support. This update preserved a unique design element while improving functionality, a theme that ran throughout the interior redesign. The dashboard layout was streamlined, with a driver-centric orientation that emphasized the car's sporting intentions.

Material innovations played a significant role in the development of second-generation TTs. Audi's use of aluminum in the ASF (Audi Space Frame) construction reduced weight while increasing rigidity. This advanced use of materials not only improved performance and efficiency but also enabled more precise manufacturing tolerances, resulting in tighter panel gaps and a higher overall quality. The mix of aluminum and high-strength steel created a structure that was both lighter and stiffer than its predecessor, enhancing both performance and safety.

The evolution of the TT's design for its second generation was a masterclass in refining a classic. By carefully updating key elements while preserving the essence of the original, Audi created a car that felt both familiar and excitingly new. The sharper lines, improved aerodynamics, refined interior, and innovative use of materials all contributed to a design that respected its past while confidently stepping into the future. This approach ensured that the second-generation TT not only lived up to the legacy of its predecessor but also set a new standard for sports car design in the 21st century.

Section 7.2: Powertrain Advancements

The second generation of the Audi TT brought significant improvements to its powertrain, offering enthusiasts a range of options that balanced performance, efficiency, and driving dynamics. At the heart of these advancements was a diverse lineup of engines, each tailored to meet different driver preferences and market demands.

The engine options for the new TT were carefully curated to provide a mix of power and efficiency. The lineup included a 2.0-liter TFSI (Turbocharged Fuel Stratified Injection) engine, which became a popular choice for its blend of performance and fuel economy. This four-cylinder powerplant delivered punchy acceleration and respectable fuel efficiency, making it an ideal choice for daily drivers and weekend enthusiasts alike. For those seeking more power, Audi offered a 3.2-liter VR6 engine. This six-cylinder unit provided a smoother power delivery and a more sonorous exhaust note, catering to drivers who prioritized performance and aural satisfaction.

One of the most significant powertrain advancements in the second-generation TT was the refinement of Audi's Quattro all-wheel-drive system. The updated Quattro system represented a leap forward in terms of capability and sophistication. It can now dynamically distribute power between the front and rear axles, sending up to 100% of the engine's torque to the rear wheels when needed. This feature not only enhanced the TT's traction in adverse weather but also improved its handling, allowing for a more engaging driving experience on twisty roads.

Transmission choices were another area where Audi made significant strides. While a six-speed manual gearbox remained available for purists, the introduction of the S tronic dual-clutch transmission was a revolution. This innovative gearbox offered lightning-fast gear changes, combining the engagement of a manual with the convenience of an automatic. The S tronic transmission could shift gears in milliseconds, keeping the engine in its optimal power band, thereby enhancing both performance and efficiency.

Performance metrics for the second-generation TT were impressive across the board. The TTS variant in particular showcased the platform's potential. Equipped with a high-output version of the 2.0 TFSI engine, the TTS could sprint from 0-60 mph in just 4.9 seconds, a figure that put it in competition with much more expensive sports cars. This level of performance was not just about straight-line speed; the TTS also offered enhanced handling and braking capabilities to match its increased power output.

Despite the focus on performance, Audi did not neglect efficiency. The 2.0 TFSI engine, in particular, offered improved fuel economy compared to its predecessor. This was achieved through advancements in engine management systems, more efficient turbocharging, and the use of direct fuel injection technology. As a result, the second-generation TT delivered increased performance without a corresponding increase in fuel consumption, a balance that resonated well with environmentally conscious sports car buyers.

The powertrain advancements in the second-generation Audi TT weren't just about raw numbers; they were about enhancing the overall driving experience. The combination of powerful, efficient engines, advanced all-wheel-drive technology, and innovative transmission options created a car that was both exhilarating to drive and practical for everyday use. This blend of performance and usability broadened the TT's appeal, attracting both hardcore enthusiasts and newcomers to the sports car segment.

In essence, the powertrain advancements of the second-generation TT embodied Audi's commitment to technological progress. By offering a range of engines to suit different tastes, refining the Quattro system for better performance and safety, and introducing cutting-edge transmission technology, Audi ensured that the new TT was not just a style icon but a driver's car. These improvements set a new benchmark in the compact sports car segment and laid the foundation for future developments in the TT lineup.

Section 7.3: Chassis and Suspension

The second-generation Audi TT introduced significant improvements to its chassis and suspension, enhancing both ride comfort and handling. At the heart of these advancements was the adoption of the MQB (Modularer Querbaukasten) platform, which is also used in other Volkswagen Group vehicles like the Golf. This platform sharing allowed Audi to leverage economies of scale in production without compromising the TT's unique character. The MQB platform provided a solid foundation for the TT, offering improved rigidity and better weight distribution compared to its predecessor.

Audi's engineers focused on refining the TT's suspension setup to deliver a more engaging driving experience. The standard suspension was tuned to provide a balance between comfort and sportiness, with revised spring rates and damper settings. However, it was the optional Magnetic Ride suspension that truly showcased Audi's commitment to cutting-edge technology.

This adaptive damping system used magnetorheological fluid to adjust suspension stiffness in real-time, responding to road conditions and driver inputs. The result was a car that could transform from a comfortable cruiser to a sharp-handling sports car at the push of a button.

The steering system also received a significant upgrade in the second-generation TT. Audi implemented an electromechanical power steering setup, replacing the hydraulic system of the previous model. This new system offered variable assistance based on vehicle speed, providing more assistance at low speeds for easier maneuvering and less at high speeds for improved feedback and precision. The electromechanical system also contributed to improved fuel efficiency by reducing parasitic losses on the engine.

Braking performance was another area of focus for Audi's engineers. The second-generation TT featured larger disc brakes all around, with ventilated discs at the front for improved heat dissipation during spirited driving. The brake calipers were also upgraded, offering better bite and pedal feel. The ABS was updated with the latest algorithms, resulting in shorter stopping distances and improved stability during hard braking. Higher-performance models like the TTS received even more substantial braking upgrades, including cross-drilled rotors for enhanced cooling.

Weight distribution played a crucial role in the TT's handling characteristics. Audi's engineers made a conscious effort to balance the car's weight more evenly between the front and rear axles. This was achieved in part by moving the engine slightly forward in the chassis compared to the previous generation. The result was a near 50:50 weight distribution in some variants, contributing to more neutral handling and improved turn-in response.

The use of lightweight materials throughout the chassis and body also played a significant role in the TT's dynamic improvements. Audi continued to utilize its Audi Space Frame (ASF) technology, incorporating aluminum and high-strength steel to reduce weight while increasing rigidity. This not only improved handling and performance but also contributed to better fuel efficiency and reduced emissions.

Audi also paid attention to the details that enhance the overall driving experience. The second-generation TT featured a more rigid body structure, which reduced cabin noise and vibration. This, combined with the refined suspension, resulted in a car that was more comfortable for long-distance cruising without sacrificing its sporty character.

The chassis and suspension enhancements of the second-generation Audi TT weren't just about numbers on a spec sheet; they fundamentally transformed the car's character. The original TT was sometimes criticized for prioritizing style over driving dynamics, but this new generation silenced those critics. It offered a level of driver engagement and capability that put it in contention with dedicated sports cars, all while maintaining the design and everyday usability that made the TT an icon.

Whether carving through mountain roads, cruising on the Autobahn, or navigating city streets, the second-generation TT's chassis and suspension setup provided a versatile and rewarding driving experience. It struck a delicate balance between comfort and performance, technology and tradition, setting a new benchmark in the sports car segment and paving the way for even more advanced developments in future generations.

Section 7.4: Technology Integration

The second-generation Audi TT marked a significant leap forward in technology integration, blending cutting-edge features with the car's signature design and performance.

This evolution reflected Audi's commitment to staying at the forefront of automotive innovation while enhancing the driving experience for TT enthusiasts.

At the heart of this technological advancement was the introduction of the optional Audi MMI (Multimedia Interface) system. This sophisticated infotainment platform brought a new level of connectivity and control to the TT's cockpit. The MMI system offered advanced navigation capabilities, enabling drivers to find their way with ease, whether on familiar roads or in new territory.

It also provided enhanced audio controls, enabling users to manage their music libraries and streaming services with intuitive ease. The integration of this system represented a significant step forward in making the TT not just a driver's car, but a technologically advanced companion for the modern motorist.

Safety and convenience were also key areas of focus in the second-generation TT's technology suite. The introduction of parking sensors and an optional rearview camera significantly improved the car's maneuverability in tight spaces. These features were particularly welcome given the TT's compact dimensions and limited rear visibility, a design characteristic.

The addition of these driver-assistance technologies made the TT more accessible to a broader range of drivers, enhancing its appeal as a daily driver without compromising its sports-car credentials. Lighting technology improved dramatically in this generation of the TT.

The optional xenon headlights with LED daytime running lights became a feature, not just improving visibility but also contributing to the car's modern aesthetic. This lighting signature would become a hallmark of Audi design, influencing the entire brand's lineup in years to come. The advanced lighting didn't just enhance the car's looks; it significantly improved night-time driving safety and visibility, a crucial factor for a performance-oriented vehicle.

Inside the cabin, climate control received a stylish and functional update. The new system featured turbine-style air vents, a design element that would become an Audi trademark. These vents were not only visually striking but also highly effective in distributing air throughout the cabin. The climate control system itself was more advanced, offering precise temperature management to ensure comfort in all driving conditions. This attention to detail in the cabin environment underscored Audi's commitment to creating a holistic driving experience that appealed to all senses.

For audiophiles, the second-generation TT offered a significant upgrade with an optional Bose surround-sound system. This high-fidelity audio setup was carefully tuned to the TT's unique cabin acoustics, providing an immersive listening experience. The system's ability to deliver crisp, clear sound even at high speeds further enhanced the TT's grand touring capabilities, making long drives as enjoyable as spirited sprints on winding roads.

The integration of these technologies into the second-generation TT went beyond adding new features. It signified a shift in the sports car paradigm, where advanced technology became an integral part of the performance and luxury experience. Audi managed to incorporate these elements without diluting the TT's core appeal as a driver-focused sports car. Instead, these technologies enhanced the car's usability and appeal, making it a more well-rounded and sophisticated vehicle.

This technological evolution in the TT also served as a preview of features that would eventually become standard across Audi's lineup and the broader automotive industry. The success of these integrations in the TT helped pave the way for even more advanced systems in future models, solidifying Audi's reputation as a leader in automotive technology.

In essence, the technology integration in the second-generation Audi TT represented a harmonious blend of performance, style, and cutting-edge features. It demonstrated that a true sports car could embrace modern technology without losing its soul, setting a new standard for what drivers could expect from a compact sports coupe in the 21st century.

Section 7.5: Model Variants

The second generation of the Audi TT introduced a diverse range of model variants, each catering to different driver preferences and market segments. This expansion of the TT lineup demonstrated Audi's commitment to broadening the appeal of their renowned sports car while maintaining its core identity.

The Audi TT: A Journey Through Design and Innovation

At the heart of the range was the TT Coupe, the standard bearer of the TT legacy. This model offered the perfect blend of style and everyday usability that had made the original TT so popular. The Coupe retained the silhouette that had become synonymous with the TT name, but with sharper lines and more aggressive styling cues. It provided a comfortable, practical option for those who wanted the TT experience without compromising daily drivability.

For those seeking open-top thrills, the TT Roadster remained a compelling option. The Roadster variant showcased Audi's engineering prowess with its fully automatic soft top, which could be operated at speeds up to 31 mph. This feature allowed drivers to respond quickly to sudden weather changes or simply indulge in spontaneous open-air driving. The Roadster's structural reinforcements ensured that it maintained the Coupe's rigidity and handling characteristics, offering a convertible experience without significant compromise.

The introduction of the TTS model marked a significant step up in the TT's performance capabilities. Boasting an uprated 2.0 TFSI engine producing 272 horsepower, the TTS bridged the gap between the standard TT and the future TT RS. It featured a host of performance enhancements, including a more aggressive suspension setup, larger brakes, and unique styling elements. The TTS catered to enthusiasts who demanded more power and sharper handling, establishing itself as the choice for drivers seeking a more focused sports-car experience.

Audi also offered several special editions throughout the second generation's lifecycle, further differentiating and adding exclusivity. One notable example was the TT S line competition package, which added exclusive exterior details and interior trims. These special editions often featured unique color combinations, bespoke alloy wheels, and interior appointments not available on standard models. They appealed to collectors and enthusiasts who sought a more TT experience.

For those who wanted to create a truly personalized TT, Audi offered an extensive range of customization options through their Audi Exclusive program. This allowed customers to specify unique paint colors, interior trims, and other bespoke details. From subtle personalization to bold statements, the Audi Exclusive program ensured that no two TTs need be precisely alike. This level of customization was particularly appealing to discerning buyers who viewed their TT not just as a mode of transport, but as a personal statement.

Each variant in the TT range, from the base Coupe to the high-performance TTS and the personalized Exclusive models, shared the same fundamental DNA that made the TT special. They all featured the styling, driver-focused interiors, and engaging driving dynamics that had become hallmarks of the TT brand. Yet each offered something unique, ensuring that there was a TT to suit a wide range of tastes and requirements.

The diversity of the TT's second-generation range was a testament to the model's versatility and broad appeal. By offering multiple variants, Audi ensured the TT could compete across various segments of the sports car market, from those seeking a stylish daily driver to enthusiasts seeking track-ready performance. This strategy not only broadened the TT's market appeal but also cemented its position as a cornerstone of Audi's performance car lineup.

Section 7.6: Critical Reception

The second-generation Audi TT's debut was met with keen interest from the automotive world, and its reception proved that Audi had successfully evolved their iconic sports car. The automotive press played a crucial role in shaping public opinion, with its reviews overwhelmingly positive.

Car and Driver, one of the most respected voices in automotive journalism, praised the TT's "sharp handling and premium feel," noting that the new model had addressed many of the criticisms leveled at its predecessor. The magazine particularly highlighted the improved driving dynamics, stating that the second-generation TT felt "more sports car than style statement." Similarly, Motor Trend commended Audi for maintaining the TT's essence while significantly enhancing its performance credentials.

In comparison tests, the new TT consistently held its own against fierce competition. In a notable face-off with the BMW Z4 and Mercedes SLK, Autocar declared the TT "the most well-rounded sports car," citing its blend of performance, comfort, and daily usability. The TT's versatility was frequently mentioned as a key strength, with reviewers appreciating its ability to serve as both a capable corner-carver and a comfortable grand tourer.

The TT's critical acclaim translated into numerous awards and accolades. Top Gear magazine named it "Coupe of the Year" in 2006, praising its design evolution and improved driving experience. The car also received the "Golden Steering Wheel" award in Germany, a testament to its appeal in the demanding European market. These accolades not only validated Audi's approach to updating the TT but also helped boost the model's profile among potential buyers.

Owner feedback played a crucial role in assessing the TT's real-world performance and appeal. TT forums buzzed with praise for the improved performance and retained design identity. Many second-generation buyers were upgrading from the first-generation model and frequently commented on noticeable improvements in handling, power delivery, and interior quality. The introduction of the S tronic dual-clutch transmission was particularly well-received, with owners lauding its quick shifts and smooth operation.

However, it wasn't all universal praise. Some purists lamented the loss of the original TT's more minimalist interior design, feeling that the new model had lost some of its predecessor's unique character in pursuit of broader appeal. Additionally, while the base model's performance was generally well-regarded, some reviewers felt that it lacked the outright sportiness of more focused competitors.

Sales figures ultimately provided concrete evidence of the second-generation TT's success. The model surpassed first-year sales projections by 15%, cementing its place in Audi's lineup and proving that the company had struck the right balance between evolution and tradition. In key markets like the UK and Germany, the TT quickly became a leader in its segment, often outperforming its main rivals from BMW and Mercedes-Benz.

The TT's strong sales performance was awe-inspiring, given the challenging economic conditions of the late 2000s. While many sports car manufacturers saw declining sales during the global financial crisis, the TT's combination of style, performance, and relative affordability helped it maintain a strong market position.

Notably, the critical and commercial success of the second-generation TT played a significant role in elevating Audi's overall brand image. The model was often featured prominently in Audi's marketing materials, serving as a halo car that embodied the brand's commitment to design excellence and technological innovation.

As the second-generation TT matured, it continued to receive updates and improvements, including the introduction of more powerful variants like the TTS. These models received their own wave of positive reviews, with many critics viewing them as the ultimate expression of the TT concept.

In retrospect, the critical reception of the second-generation Audi TT validates the challenging task Audi set for itself: improving an icon. By carefully evolving the TT's design while significantly enhancing its performance and technology, Audi created a sports car that not only lived up to its predecessor's legacy but in many ways surpassed it, setting a new benchmark for the model and the segment as a whole.

Section 7.7: Legacy and Influence

The second-generation Audi TT left an indelible mark on the automotive landscape, extending far beyond its own model range. Its impact on the Audi brand was profound, solidifying the company's reputation for cutting-edge design and performance. The TT's success bolstered Audi's image as a manufacturer of premium sports cars, elevating the entire brand in the eyes of consumers and enthusiasts alike.

The influence of the second-generation TT on competitors was equally significant. As rival manufacturers scrambled to match the TT's blend of style, performance, and technology, the sports car segment as a whole evolved. Following the TT's lead, competitors began offering dual-clutch transmissions in their sports cars, recognizing the demand for lightning-fast gear changes and improved efficiency. The TT's success also prompted other manufacturers to pay closer attention to interior design and material quality, raising the bar for the entire segment.

One of the most lasting legacies of the second-generation TT was its role as a technological pioneer for the Audi brand. Many innovations introduced or refined in the TT would later find their way into other Audi models. The virtual cockpit concept, which debuted in the TT, would later be adopted across Audi's range, revolutionizing how drivers interact with their vehicles. This trickle-down effect of technology showcased the TT's importance as a halo product for the brand.

The second-generation TT also made its mark in motorsport, proving that its performance credentials were more than just marketing hype. A modified TTS set a new class record at the Pikes Peak Hill Climb in 2009, demonstrating the platform's potential when pushed to its limits. This achievement not only boosted the TT's credibility among performance enthusiasts but also provided valuable data and experience that Audi could apply to future road and race cars.

As the second-generation TT ages, its standing among car collectors continues to grow. Low-mileage TTS models, in particular, have begun to appreciate, signaling the car's transition from merely a used sports car to a future classic. This appreciation is a testament to the enduring appeal of the TT's design and the respect it commands among automotive enthusiasts.

The second-generation TT's legacy is one of successful evolution. It proved that the design could be updated and improved without losing its essential character. By refining the original concept with enhanced performance, advanced technology, and improved quality, Audi created a car that not only lived up to its predecessor's reputation but surpassed it in many ways.

Ultimately, the second-generation TT served as a bridge between the original concept car-inspired first generation and the increasingly high-tech and performance-focused later models. It maintained the TT's position as a design icon while significantly enhancing its credentials as a bona fide sports car. This delicate balance between honoring the past and embracing the future would continue to define the TT through subsequent generations, ensuring its place as one of Audi's most beloved and influential models.

The Audi TT: A Journey Through Design and Innovation

Chapter 8: Pushing Boundaries: TT RS and High-Performance Variants

Section 8.1: The Birth of the TT RS

The Audi TT RS represents the pinnacle of performance in the TT lineup, born from a desire to create a sports car that could compete with the best in its class. The concept behind the TT RS was ambitious: to take the already popular TT and transform it into a high-performance machine that would capture the hearts of driving enthusiasts worldwide.

The development process of the TT RS was not without its challenges. Audi's engineers faced the daunting task of significantly boosting performance while maintaining the TT's signature design and everyday usability. This balancing act required countless hours of research, testing, and refinement. The team had to push the boundaries of engine technology, aerodynamics, and chassis design to achieve their lofty goals.

Key figures in the TT RS's development included Michael Dick, Audi's Board Member for Technical Development at the time, and Stephan Reil, who headed Audi's high-performance quattro GmbH division. These visionaries, along with a dedicated team of engineers and designers, worked tirelessly to bring the TT RS to life. Their expertise and passion were instrumental in creating a car that would exceed expectations and set new standards in the sports car segment.

Within Audi's lineup, the TT RS was positioned as the halo model of the TT range, designed to showcase the brand's performance capabilities and quattro all-wheel-drive technology. It was intended to appeal to drivers who demanded the ultimate in performance from their sports car, while still retaining the practicality and style that made the TT so popular. The goal was clear: to create a vehicle that could compete with established players in the high-performance sports car market, such as Porsche's Cayman S.

When Audi first announced the TT RS, the automotive world took notice. Initial reactions from both the public and industry insiders were overwhelmingly positive. The prospect of a high-performance TT with a unique five-cylinder engine generated considerable buzz. Enthusiasts eagerly anticipated the chance to experience the TT RS's promised blend of style and performance. At the same time, competitors undoubtedly felt a twinge of concern at the arrival of this potent new rival.

Automotive journalists praised Audi's bold move, recognizing the TT RS's potential to shake up the sports car segment. Many saw it as a return to form for Audi, harkening back to the brand's illustrious racing heritage and the legendary Sport Quattro of the 1980s. The announcement of the TT RS was a clear statement of intent from Audi, signaling its commitment to performance and determination to push the boundaries of what was possible in a compact sports car.

As anticipation for the TT RS's release built, it became clear that Audi had created something special. The birth of the TT RS marked a new chapter in the TT's history, one that would redefine the model's capabilities and cement its status as a performance icon.

Section 8.2: Engine and Powertrain Innovations

At the heart of the TT RS lies its aggressive powerplant, a testament to Audi's engineering prowess and commitment to performance. The centerpiece of this high-performance variant is the 2.5-liter turbocharged five-cylinder engine, a powerhouse that pays homage to Audi's illustrious racing heritage while embracing modern technology.

This engine is a marvel of compact design and efficient power delivery. Producing an impressive 394 horsepower and 354 lb-ft of torque in its latest iteration, it catapults the TT RS into supercar territory. The unique five-cylinder configuration provides a distinctive engine note, a throaty growl that sets the TT RS apart from its four and six-cylinder competitors. Audi's engineers have employed advanced materials, including an aluminum crankcase, to reduce weight and improve overall performance.

To harness this impressive power, Audi enhanced its quattro all-wheel-drive system specifically for the TT RS. This updated system can variably distribute torque between the front and rear axles, sending up to 100% of the power to the rear wheels when needed. This dynamic torque distribution not only improves traction in various driving conditions but also enhances the car's agility and cornering capabilities.

The TT RS offers a sophisticated seven-speed S tronic dual-clutch transmission, a perfect match for the high-revving five-cylinder engine. This gearbox delivers lightning-fast shifts, whether in automatic mode or with the steering wheel-mounted paddles. The transmission's programming is tailored to maximize the engine's power band, ensuring that the TT RS is always in the optimal gear for explosive acceleration or efficient cruising.

When compared to its competitors, the TT RS engine stands out for its unique configuration and power density. While many rivals opt for turbocharged four-cylinder or naturally aspirated six-cylinder engines, Audi's five-cylinder approach offers a compelling blend of compact packaging, smooth power delivery, and a character that sets it apart in the sports car segment.

The result of these powertrain innovations is a car that delivers not only blistering straight-line performance but also an engaging, visceral driving experience. The TT RS can sprint from 0 to 60 mph in just 3.6 seconds, a figure that puts it in the company of much more expensive supercars. Yet, it's not just about raw speed; the linear power delivery and responsive nature of the engine make the TT RS a joy to drive in a variety of situations, from winding mountain roads to high-speed autobahn runs.

Audi's commitment to continuous improvement is evident in the evolution of the TT RS powertrain. Each iteration has seen incremental enhancements in power output, efficiency, and responsiveness. This dedication to refining the formula ensures that the TT RS remains at the forefront of compact sports-car performance, challenging perceptions of what's possible from a small-displacement engine.

The engine and powertrain innovations in the TT RS represent a perfect synthesis of Audi's racing heritage and cutting-edge technology. They not only elevate the TT's performance to new heights but also showcase Audi's ability to create emotionally engaging, high-performance machines that resonate with driving enthusiasts worldwide.

Section 8.3: Chassis and Suspension Upgrades

The Audi TT RS didn't just receive a heart transplant with its powerful five-cylinder engine; it also underwent significant chassis and suspension upgrades to handle the increased power and deliver a truly exhilarating driving experience.

At the core of these improvements was a reinforced chassis design. Audi's engineers meticulously strengthened key areas of the TT's already rigid structure, focusing on critical stress points to enhance overall torsional rigidity. This reinforcement not only improved handling characteristics but also enhanced safety performance. The stiffer chassis allowed for more precise tuning of the suspension components, resulting in a more responsive and predictable car at the limits of adhesion.

The sport-tuned suspension system was a crucial element in transforming the TT into a true high-performance machine. The TT RS featured a lower ride height than the standard TT, reducing the center of gravity and minimizing body roll during aggressive cornering. Stiffer springs and dampers were employed to provide sharper turn-in response and better body control.

Audi also offered an optional magnetic ride suspension system that used magnetorheological fluid in the dampers to allow real-time adjustments to suspension characteristics. This technology enabled drivers to switch between comfort and dynamic modes, adapting the car's behavior to different driving conditions with the push of a button.

Braking performance saw a significant upgrade to match the TT RS's increased speed potential. The standard brake system featured larger rotors and more robust calipers, providing excellent stopping power and fade resistance during high-performance driving. For those seeking the ultimate in braking performance, Audi offered optional carbon-ceramic brakes.

These not only provided even better heat dissipation and reduced brake fade but also contributed to lower unsprung weight, further enhancing the car's agility.

Aerodynamic enhancements played a crucial role in improving the TT RS's handling at high speeds. A redesigned front splitter reduced lift at the front axle, while the prominent rear wing generated significant downforce at the rear. These aerodynamic elements worked in concert to keep the car planted during high-speed cornering and improve overall stability. Side skirts and a rear diffuser completed the aerodynamic package, optimizing airflow around and under the vehicle.

Weight reduction was another key focus area for Audi's engineers. Despite adding performance components, they managed to keep the TT RS's weight in check by using lightweight materials. The hood and various body panels were crafted from aluminum, while high-strength steel was strategically used throughout the chassis. Even the wheels were designed with weight savings in mind, with Audi offering forged aluminum options that reduced unsprung mass and improved the car's responsiveness to steering inputs.

The cumulative effect of these chassis and suspension upgrades was a car that felt alive and eager in the driver's hands. The TT RS exhibited razor-sharp handling, with minimal body roll and exceptional grip through corners. The improved weight distribution and lowered center of gravity resulted in a car that changed direction with agility and precision. Whether on a winding mountain road or a racetrack, the TT RS's chassis and suspension upgrades allowed drivers to explore the limits of performance with confidence, transforming the already capable TT into a driver's car that could compete with some of the most renowned sports cars in its class.

Section 8.4: Exterior Design Elements

The Audi TT RS's exterior design is a testament to the perfect fusion of form and function, taking the iconic TT silhouette to new heights of aggression and performance-oriented aesthetics. At the forefront of this visual transformation is the TT RS's imposing front fascia. The signature Audi Singleframe grille adopts a more menacing appearance, featuring a honeycomb pattern and an RS badge. This broader, more angular grille is flanked by large air intakes that not only enhance the car's aggressive stance but also help cool the high-performance engine.

The TT RS sets itself apart with a unique array of wheel options that perfectly complement its high-performance nature. Buyers can choose from various designs, typically ranging from 19 to 20 inches in diameter. These wheels are not just about looks; they're engineered to be lightweight yet strong, enhancing the car's handling dynamics and visual appeal. The wheel designs often feature intricate spoke patterns and can be finished in colors like matte titanium or gloss black, adding to the car's premium and sporty character.

One of the most exterior features of the TT RS is its fixed rear wing. Unlike the standard TT's retractable spoiler, the RS variant sports a prominent, fixed wing that makes a bold statement about the car's performance intentions. This isn't just for show - the wing generates significant downforce at high speeds, improving stability and cornering capabilities. The wing's design is carefully integrated into the TT's flowing lines, enhancing rather than disrupting the overall aesthetic.

Audi offers the TT RS in a palette of colors that emphasize its sporting pretensions. While classic options like Nardo Grey and Daytona Grey remain popular choices, more vibrant hues such as Catalunya Red and Kyalami Green allow owners to make a bolder statement.

These color options are often complemented by RS-specific trim packages that include black optic elements or carbon fiber accents, further distinguishing the RS from its standard TT counterparts. Compared to the standard TT, the RS variant's design elements create a noticeably more aggressive, purposeful appearance. The front end sits lower and wider, the side sills are more pronounced, and the rear diffuser is more prominent, housing larger exhaust outlets. These changes not only signal the car's increased performance capabilities but also improve its aerodynamic efficiency.

Every exterior modification on the TT RS serves a dual purpose - enhancing both aesthetics and performance. From the redesigned front bumper, which channels air more efficiently, to the subtle wheel arch extensions that accommodate wider tires, each element contributes to the car's overall dynamic performance while also creating a visually striking presence on the road.

The TT RS's exterior design successfully walks the fine line between the TT's classic, timeless shape and a more aggressive, performance-oriented aesthetic. It honors the original TT's design heritage while clearly communicating its elevated performance, resulting in a car that looks fast even when standing still.

Section 8.5: Interior Features and Technology

The Audi TT RS doesn't just excel in performance; it also offers a cabin that perfectly balances luxury, sportiness, and cutting-edge technology. Upon entering the TT RS, drivers are immediately greeted by a race-inspired cockpit that sets the tone for the exhilarating driving experience ahead.

At the heart of the TT RS's interior is the RS-specific steering wheel. This flat-bottomed, leather-wrapped masterpiece is more than just a control interface; it's a statement of intent. With its chunky grip, contrast stitching, and prominent RS badging, the wheel provides both tactile pleasure and visual appeal. The wheel also houses convenient controls for various vehicle functions, allowing drivers to adjust settings without taking their hands off the wheel – a crucial feature when pushing the car to its limits.

The sport seats in the TT RS are a testament to Audi's commitment to both comfort and performance. Available in fine Nappa leather with diamond stitching, these seats offer exceptional lateral support to keep drivers firmly in place during high-speed cornering. The seats are not only functional but also a visual treat, with RS embossing and available contrast stitching that adds to the cabin's sporty aesthetic. For those seeking the ultimate racing experience, Audi offers optional RS sport seats with even more aggressive bolstering and integrated headrests.

Audi's commitment to technological innovation shines through in the TT RS's infotainment system and driver displays. The crown jewel is the Audi Virtual Cockpit, a fully digital 12.3-inch instrument cluster that replaces traditional analog gauges. This high-resolution display can be customized to show a variety of information, from navigation maps to performance data, all directly in the driver's line of sight. The RS-specific version of the Virtual Cockpit includes unique displays such as a central tachometer with integrated speed display, reminiscent of racing cars.

For performance enthusiasts, the TT RS offers a wealth of real-time data at their fingertips. The Virtual Cockpit can display a range of RS-specific information, including tire pressure, torque, and g-force. An imposing feature is the lap timer function, which allows drivers to record lap times and analyze their performance on the track. This blend of performance data and digital display technology truly bridges the gap between road car and race car.

The audio experience in the TT RS is equally impressive, designed to complement the five-cylinder engine's exhilarating soundtrack. While the standard sound system offers excellent quality, true audiophiles can opt for the Bang & Olufsen Sound System. This premium option features 12 speakers and 680 watts of power, delivering concert-hall quality sound that can still be appreciated even with the engine roaring at full throttle.

Audi has also paid attention to the finer details that enhance the overall interior ambiance. The air vents, a feature in all TT models, receive special treatment in the RS variant. With their turbine-inspired design and integrated climate controls, they're both functional and aesthetically pleasing. Aluminum race inlays and carbon fiber trim options further reinforce the car's performance pedigree.

Lighting plays a crucial role in setting the mood inside the TT RS. The standard LED interior lighting package bathes the cabin in a soft glow. In contrast, the available extended LED interior lighting package allows for personalized color schemes, adding another layer of customization to the driving experience.

In essence, the TT RS's interior is a perfect reflection of the car's dual nature. It offers the luxury and technology expected of a premium Audi product, while simultaneously providing the focused, driver-centric environment of a high-performance sports car. This careful balance ensures that whether you're attacking a racetrack or cruising on the highway, the TT RS's interior is always the perfect place to be.

Section 8.6: Performance Benchmarks and Testing

The Audi TT RS's performance capabilities are not just marketing hype; impressive real-world results back them. When it comes to acceleration, the TT RS is a force to be reckoned with. The latest model can sprint from 0 to 60 mph in a blistering 3.6 seconds, putting it in the same league as many supercars. This incredible acceleration is made possible by the potent combination of the 2.5-liter turbocharged engine, quattro all-wheel-drive system, and lightning-fast dual-clutch transmission.

Quarter-mile times further demonstrate the TT RS's prowess. It consistently clocks in at around 11.9 seconds, crossing the finish line at speeds approaching 120 mph. These figures aren't just impressive for a compact sports car; they're competitive with vehicles costing twice as much.

While the TT RS is electronically limited to 155 mph in standard form, Audi offers an option to increase this to 174 mph for those who genuinely want to push the envelope. It's worth noting that without the limiter, the car can reach even higher speeds, a testament to its aerodynamic efficiency and powerful engine.

Straight-line speed is only part of the equation, however. The TT RS truly shines in handling and cornering. In skidpad tests, which measure a car's ability to maintain grip while cornering, the TT RS consistently achieves over 1.0 g of lateral acceleration. This exceptional grip, combined with the car's responsive steering and well-tuned suspension, translates to incredible agility on twisty roads and race tracks alike.

Speaking of race tracks, the TT RS has proven its mettle on the world's most challenging circuit: the Nürburgring Nordschleife. Audi's test drivers have lapped the grueling 12.9-mile track in just 7 minutes and 40 seconds, a time that puts the TT RS in elite company. This lap time isn't just impressive for a compact sports car; it's competitive with dedicated sports cars and even some supercars.

Automotive journalists have consistently praised the TT RS for its real-world performance. Many have noted that the car feels even quicker than its official numbers suggest, thanks to its immediate response and the soundtrack from its unique five-cylinder engine. The quattro all-wheel-drive system has received particular acclaim for its ability to put power down effectively in all conditions, making the TT RS an actual all-weather performance car.

Braking performance is another area where the TT RS excels. In 70-0 mph brake tests, the car consistently stops in distances under 150 feet, a figure that rivals many supercars. This stellar braking performance, combined with the car's nimble handling, gives drivers the confidence to push harder and brake later when attacking corners.

It's not just the raw performance figures that impress, but how accessible the TT RS makes this performance. Unlike some high-strung sports cars that require a skilled hand to extract their full potential, the TT RS delivers its performance in a manner that's approachable for drivers of varying skill levels. This combination of extreme capability and everyday usability is the TT RS's most impressive feat.

In the realm of compact sports cars, the Audi TT RS stands out as a true giant-killer. Its performance benchmarks and real-world testing results prove that it's not just a stylish sports coupe, but a genuine high-performance machine capable of challenging cars from much more expensive segments. The TT RS represents the pinnacle of what Audi's engineers can achieve by pushing the boundaries of the TT platform, delivering supercar-level thrills in a compact, stylish package.

Section 8.7: Other High-Performance TT Variants

While the TT RS stands as the pinnacle of performance in the TT lineup, Audi has introduced several other high-performance variants for enthusiasts seeking elevated driving experiences.

These models bridge the gap between the standard TT and the range-topping RS, offering a spectrum of performance options for discerning drivers. The TTS serves as a middle-ground performance option, striking a balance between everyday usability and track-ready capabilities. Introduced alongside the second-generation TT, the TTS features a more powerful engine, sport-tuned suspension, and styling cues. Its 2.0-liter turbocharged four-cylinder engine delivers a significant boost in power over the base TT, while still maintaining reasonable fuel efficiency. The TTS has become a popular choice for those seeking an exhilarating driving experience without the full-on intensity of the RS model.

Audi has also released several limited edition models that showcase the TT's performance potential. A prime example is the TT RS 40 Years of quattro, unveiled in 2020 to celebrate four decades of Audi's renowned all-wheel-drive system. Limited to just 40 units worldwide, this exclusive model features unique aesthetic elements inspired by the Audi Sport quattro S1 rally car, including a striking Alpine White paint job with decal accents. Performance enhancements include an upgraded exhaust system and an increased top speed of 174 mph, making it a true collector's item for Audi enthusiasts.

The TT RS Plus, introduced in 2012, pushed the TT's performance even further. This enhanced version of the already potent TT RS boasted a power increase to 360 horsepower, achieved through engine tuning and an upgraded exhaust system. The Plus model also featured unique design elements, such as a carbon fiber engine cover and special alloy wheels, setting it apart from the standard TT RS.

Throughout the TT's history, Audi has also tantalized enthusiasts with concept and prototype high-performance models. These exercises in automotive design and engineering have showcased potential future directions for the TT lineup.

Notable examples include the TT ultra quattro concept, which demonstrated extreme weight reduction techniques, and the TT sport turbo concept, featuring an electrically driven compressor for enhanced performance.

Looking ahead, rumors and industry speculation suggest that Audi may be developing even more potent performance variants of the TT. With the automotive industry's shift towards electrification, there's potential for hybrid or fully electric high-performance TT models. These could leverage instant torque delivery and advanced battery technology to achieve new levels of acceleration and handling prowess.

As Audi continues to push the boundaries of performance and technology, the TT lineup remains a fertile ground for innovation. Whether through limited-edition models, performance-focused variants, or cutting-edge prototypes, the brand consistently demonstrates its commitment to evolving the TT beyond its sports-car roots. These high-performance variants not only cater to enthusiasts but also serve as technological showcases, often previewing features and capabilities that eventually trickle down to other models in the Audi range.

The Audi TT: A Journey Through Design and Innovation

Chapter 9: The Third Generation: Embracing Modern Technology

Section 9.1: Design Evolution

The third-generation Audi TT, unveiled in 2014, marked a significant evolution in the celebrated sports car's design. While maintaining its unmistakable silhouette, the new TT adopted a more angular, aggressive aesthetic, aligning with Audi's contemporary design language.

The exterior styling changes were immediately apparent, with sharper lines and more pronounced features giving the TT a bolder presence on the road. The front grille became more hexagonal and pronounced, lending the car a more assertive stance. This redesigned fascia was complemented by sleeker headlights featuring Audi's signature LED daytime running lights, enhancing both the car's visual appeal and visibility.

Aerodynamic improvements played a crucial role in the third-generation TT's design. Every curve and contour was meticulously crafted to reduce drag and improve high-speed stability. The redesigned rear spoiler, for instance, now automatically deploys at speeds over 75 mph, reducing lift and enhancing the car's performance. This active aerodynamic element not only improved the TT's handling characteristics but also added a touch of theatrical flair to its already eye-catching design.

Audi's commitment to lightweight construction was evident in the new TT. The extensive use of aluminum and high-strength steel resulted in a car that was up to 50 kg lighter than its predecessor. This weight reduction not only improved performance and fuel efficiency but also enhanced the car's agility and responsiveness. The intelligent mix of materials showcased Audi's expertise in advanced manufacturing techniques, setting a new standard for sports car construction.

Lighting technology took a significant leap forward in the third-generation TT. It was one of the first Audi models to offer optional Matrix LED headlights, a truly amazing system that provided superior illumination without blinding oncoming drivers. These adaptive headlights could selectively dim individual LEDs to create a constantly adapting light pattern, maximizing visibility while ensuring safety for other road users. This technology not only enhanced the TT's functionality but also gave it a futuristic appearance, especially at night.

The wheels and tires of the new TT were also carefully considered in its design evolution. Audi offered a range of wheel designs from 17 to 20 inches in diameter, allowing owners to personalize their car's look. These wheels were not just about aesthetics; they were designed to optimize the balance between performance and comfort. Specially developed tires complement these wheels, engineered to enhance grip, reduce road noise, and improve fuel efficiency. The larger wheel options, in particular, filled out the TT's wheel arches perfectly, contributing to its athletic stance.

In every aspect of its exterior design, the third-generation Audi TT demonstrated a perfect blend of form and function. It respected the car's heritage while pushing the boundaries of style, aerodynamics, and technology. The result was a sports car that looked thoroughly modern and dynamic, yet was instantly recognizable as a TT. This evolution in design not only pleased longtime fans of the model but also attracted a new generation of enthusiasts, ensuring the TT's continued relevance in an ever-changing automotive landscape.

Section 9.2: Powertrain Advancements

The third-generation Audi TT brought significant advancements to its powertrain, offering a range of engines and transmissions that pushed the boundaries of performance and efficiency. At the heart of the lineup was a 2.0-liter TFSI engine, a testament to Audi's expertise in turbocharged technology. This versatile powerplant was offered in various states of tune, with outputs ranging from a respectable 230 horsepower in the base model to an impressive 310 horsepower in the high-performance TTS variant. The engine's broad power band and quick-spooling turbocharger provided drivers with responsive acceleration across the rev range, making the TT equally at home in urban traffic and on winding country roads.

Complementing the engine's capabilities was Audi's new seven-speed S tronic dual-clutch transmission. This gearbox represented a significant leap forward in transmission technology, offering lightning-fast gear changes that were imperceptible to the driver. The S tronic's ability to pre-select the next gear before it was needed resulted in seamless power delivery and improved fuel efficiency. For drivers who preferred a more engaging experience, a six-speed manual transmission was also available on select models, offering precise gear selection and a more direct connection to the drivetrain.

The third-generation TT also saw updates to Audi's renowned Quattro all-wheel-drive system. This latest iteration of Quattro can now dynamically distribute torque between the front and rear axles, sending up to 100% of power to either end as needed. This not only enhanced traction in adverse weather conditions but also improved the car's handling dynamics. The system could detect wheel slip and adjust power distribution in milliseconds, providing a level of surefootedness that belied the TT's sports car nature.

Despite the increase in performance, Audi engineers improved the TT's fuel efficiency. Through the implementation of start-stop technology, which automatically shuts off the engine when the car is stationary, and improved engine management systems, the new TT achieved up to 14% better fuel economy than its predecessor. This improvement was particularly notable given the car's increased power output, demonstrating Audi's commitment to balancing performance with environmental responsibility.

The third-generation TT's performance figures were impressive across the board. The standard model could accelerate from 0-60 mph in around 5.3 seconds, while the more potent TTS shaved that time down to just 4.6 seconds, a half-second improvement over the previous generation. These figures put the TT in direct competition with some of the most renowned sports cars in its class, yet it achieved this performance while maintaining the practicality and comfort for which the model had become known.

Audi also introduced a range of driving modes through its Drive Select system, allowing drivers to tailor the car's performance characteristics to their preferences and driving conditions. These modes affected various aspects of the powertrain, including throttle response, shift points for the S tronic transmission, and even the engine sound. In Dynamic mode, for instance, the TT would hold gears longer, sharpen throttle response, and produce a more aggressive exhaust note, transforming the car's character from a comfortable cruiser to a focused sports car at the touch of a button.

The powertrain advancements in the third-generation Audi TT represented a significant step forward for the model. By combining increased performance with improved efficiency and offering a level of customization through its various driving modes, Audi created a powertrain that could satisfy both enthusiast drivers and those seeking a more relaxed driving experience. These improvements ensured that the TT remained competitive in an increasingly crowded sports car market, while also pointing the way forward for future Audi models in terms of performance and efficiency.

Section 9.3: Interior Technology

The third-generation Audi TT's interior marked a significant leap forward in automotive technology, seamlessly blending cutting-edge features with the car's signature design language. At the heart of this technological revolution was the introduction of the Virtual Cockpit, a fully digital instrument cluster that transformed how drivers interact with their vehicle.

The 12.3-inch high-resolution display of the Virtual Cockpit replaced traditional analog gauges, offering a customizable interface that could prioritize different information based on the driver's preferences. With the ability to switch between classic view, infotainment mode, and a sport display, the Virtual Cockpit offered unprecedented flexibility. Drivers could now view full-screen navigation maps, audio information, or performance data without taking their eyes off the road ahead.

Complementing the Virtual Cockpit was Audi's latest MMI (Multi Media Interface) system. This new iteration featured natural-language voice control, allowing drivers to interact with the system using conversational commands rather than rigid, predefined phrases. The system's handwriting recognition feature, integrated into the top of the rotary controller, made inputting destinations or searching for contacts a breeze, even while on the move.

Connectivity took center stage in the third-generation TT with the introduction of Audi connect. This suite of online services transformed the car into a mobile Wi-Fi hotspot, enabling passengers to stay connected on the go. Moreover, it provided real-time information services, including traffic updates, weather forecasts, and social media integration, keeping occupants informed and entertained throughout their journey.

For audiophiles, the TT offered a range of sophisticated sound systems, culminating in the optional Bang & Olufsen setup. This premium audio system featured 12 speakers and a 680-watt amplifier, delivering an immersive, concert-hall-like experience. The system's advanced sound processing could adapt to ambient noise, ensuring optimal audio quality regardless of driving conditions.

One of the most visually striking and innovative features of the new interior was the climate control system. In a bold design move, Audi integrated the controls directly into the air vents themselves. This not only reduced dashboard clutter but also created a futuristic, minimalist aesthetic that perfectly complemented the TT's exterior design. The temperature, fan speed, and air distribution could all be adjusted by manipulating the vents' centers, with digital displays showing the current settings.

The technological advancements in third-generation TTs' interiors weren't just about adding features; they fundamentally changed how drivers interacted with their vehicles. By consolidating most functions into the Virtual Cockpit and streamlining the center console, Audi created a driver-centric cockpit that minimized distractions and enhanced the overall driving experience.

This tech-forward approach didn't compromise the TT's sporty character. Instead, it enhanced it, allowing drivers to access performance data more easily and customize their driving environment to suit their preferences. The result was an interior that felt both futuristic and familiar, maintaining the TT's reputation for avant-garde design while pushing the boundaries of in-car technology.

The third-generation TT's interior technology set a new benchmark in the sports car segment, influencing not only future Audi models but the broader automotive industry. It demonstrated that high-tech features could be seamlessly integrated into a driver-focused sports car, enhancing rather than detracting from the pure driving experience that TT enthusiasts had come to love.

Section 9.4: Safety Technologies

The third-generation Audi TT not only embraced performance and technology but also made significant strides in safety, bringing luxury-car-level protection to the sports-car segment. This commitment to safety was evident in the array of advanced driver assistance systems introduced in this iteration of the TT.

One of the standout features was the active lane assist system, which used a forward-facing camera to detect lane markings and gently steer the car back into its lane if it began to drift. This technology, previously found only in larger luxury sedans, was a breakthrough for the sports car market. Complementing this was the traffic sign recognition system, which used the same camera to read speed limit signs and other road warnings, displaying them prominently in the Virtual Cockpit to keep the driver informed.

Parking, often a challenge in tight urban environments, became significantly easier with the introduction of an advanced park assist system. This technology could not only identify suitable parking spaces but also automatically steer the car into both parallel and perpendicular spots. The driver only needed to control the accelerator and brake, making parking in tight spaces a breeze.

Passive safety systems also saw considerable improvements in the third-generation TT. A new head airbag system was introduced, deploying from the A-pillar to provide enhanced protection in the event of a side impact. This addition significantly improved occupant safety, especially considering the TT's compact size. The structural integrity of the TT was enhanced by using hot-formed steel in critical areas of the body structure. This high-strength material improved crash performance without adding excess weight, maintaining the car's agile handling characteristics while providing a robust safety cell for occupants.

Audi also paid attention to pedestrian safety, an increasingly important consideration in modern car design. The hood of the third-generation TT featured pyrotechnic actuators that would raise it slightly in the event of a pedestrian impact. This clever system increased the deformation zone between the hood and the rigid components underneath, potentially reducing the severity of injuries in pedestrian collisions.

The braking system also saw significant upgrades, with larger brake discs and more rigid calipers improving stopping power and brake feel. This enhancement not only contributed to the car's performance credentials but also played a crucial role in accident avoidance. Audi's pre sense system, which prepares the vehicle for an impending collision by tightening seatbelts, closing windows, and priming the brakes, was also available on the TT for the first time. This proactive approach to safety demonstrated Audi's commitment to protecting occupants even before an impact occurs.

Visibility, a crucial safety aspect, was improved with the introduction of Matrix LED headlights as an optional feature. These advanced lights could selectively dim portions of the high beam to avoid dazzling oncoming drivers while maintaining maximum illumination elsewhere, significantly enhancing nighttime driving safety.

The integration of these safety technologies in the third-generation TT represented a significant leap forward. It demonstrated that high levels of safety could be seamlessly incorporated into a sports car without compromising its performance or aesthetic appeal. By bringing these advanced safety features to the TT, Audi not only enhanced its occupants' protection but also set a new standard for safety in the sports car segment, challenging the notion that compact sports cars must compromise on safety.

Section 9.5: Special Editions and Variants

The third-generation Audi TT introduced several special editions and variants that catered to different segments of the sports car market, from high performance enthusiasts to collectors seeking exclusive models. At the top of the performance hierarchy stood the TTS, which boasted an impressive 310 horsepower and a host of performance upgrades. This variant featured larger brakes for improved stopping power and a more aggressive suspension tune, allowing drivers to fully exploit the car's capabilities on both road and track. The TTS quickly became a favorite among enthusiasts who sought a perfect balance between daily usability and exhilarating performance. rs

However, for those craving even more power and exclusivity, Audi unveiled the range-topping TT RS in 2016. This model was a true testament to Audi's commitment to performance and its rich rallying heritage. The TT RS featured a turbocharged five-cylinder engine producing a staggering 400 horsepower, a nod to the legendary Audi Quattro of the 1980s. With its engine note and blistering acceleration, the TT RS offered supercar-like performance in a compact and stylish package.

To celebrate two decades of the TT's stylish design, Audi introduced the 20 Years Edition in 2018. This special model paid homage to the original TT concept car, featuring unique paint options that included Arrow Gray and Nano Gray. Special badging, both inside and out, reminded onlookers of the TT's lasting impact on automotive design. The interior was equally exceptional, with brown Nappa leather seats featuring diamond quilting and yellow contrast stitching, echoing the baseball-stitched leather seats of the original TT. This limited production run quickly became a collector's item, appealing to longtime TT enthusiasts and automotive history buffs alike.

Audi also experimented with limited-run models and concept cars based on the third-generation TT platform. One notable example was the TT ultra quattro concept, which showcased extreme weight-saving measures. This concept car pointed towards future lightweight technologies that could be implemented in production models. While it never made it to production, the ultra quattro concept demonstrated Audi's commitment to pushing the boundaries of performance and efficiency.

The third-generation TT range wouldn't be complete without the Roadster variants. The TT Roadster featured a fabric roof that could open or close in just 10 seconds at speeds up to 31 mph, allowing drivers to quickly adapt to changing weather conditions or simply enjoy open-top motoring at a moment's notice. The Roadster retained the coupe's sharp handling characteristics while offering an even more immersive driving experience. It was available in standard, TTS, and eventually TT RS guises, catering to a wide range of preferences and performance requirements.

Each of these special editions and variants contributed to the third-generation TT's appeal, ensuring that there was a version of the car to suit almost every taste and requirement within the sports car segment. From the daily-drivable base model to the track-focused TT RS, and from the nostalgic 20 Years Edition to the versatile Roadster, the TT lineup showcased Audi's ability to spin numerous compelling variations from a single, strong foundation.

These diverse offerings helped maintain interest in the TT throughout its third generation, cementing its position as a versatile and enduring icon in the automotive world.

Section 9.6: Reception and Impact

The third-generation Audi TT made a significant splash in the automotive world, attracting the attention of critics, enthusiasts, and consumers alike. Upon its release, automotive journalists were quick to praise the new TT for its successful blend of legendary design and cutting-edge technology. Top Gear magazine, for instance, lauded the car as "a tech-fest wrapped in a tantalizingly stylish body," encapsulating the general sentiment among reviewers. The innovative Virtual Cockpit, in particular, received widespread acclaim, with many hailing it as a revolution in automotive interior design.

Despite a generally declining sports car market, the third-gen TT maintained strong sales, particularly in European markets. This success was primarily attributed to the car's ability to appeal to a wide range of buyers, from tech enthusiasts to driving purists. In Germany, the TT's home market, it consistently outperformed many of its direct competitors in the compact sports car segment.

The TT's innovations didn't go unrecognized in the industry. The Virtual Cockpit technology, which debuted in the TT, won the prestigious 'Technology of the Year' award from What Car? magazine in 2015. This accolade underscored Audi's commitment to pushing the boundaries of in-car technology and set a new standard for digital interfaces in automobiles. One of the most significant impacts of the third-generation TT was its influence on other Audi models. The Virtual Cockpit technology, having proved its worth in the TT, quickly spread to other vehicles in Audi's lineup.

It became a signature feature of the brand, enhancing the perceived value and technological prowess of Audi's entire range. This trickle-down effect demonstrated how the TT, despite being a niche sports car, played a crucial role in shaping the broader Audi brand identity. When compared to its competitors, the third-gen TT carved out a unique position in the sports car market. While rivals like the Porsche Cayman might have offered a more hardcore driving experience, the TT distinguished itself with its blend of performance, technology, and daily usability. This combination appealed to buyers who wanted a car that could deliver excitement on weekend drives while also serving as a comfortable, feature-rich daily driver.

The car's success in balancing these often-conflicting priorities was reflected in its customer satisfaction scores, which consistently ranked among the highest in its class. Owners frequently cited the car's versatility, advanced features, and design as key factors in their purchasing decision and subsequent satisfaction.

However, it's worth noting that the third-gen TT wasn't without its critics. Some purists lamented the increased focus on technology, arguing that it detracted from the pure driving experience that earlier TTs had offered. Others pointed out that while the performance had undoubtedly improved, the TT still couldn't quite match the most focused sports cars in terms of outright driving dynamics. Despite these criticisms, the third-generation Audi TT received an overwhelmingly positive reception.

It successfully modernized the TT concept for a new era, proving that this design icon could evolve with the times while maintaining its core appeal. The car's impact extended beyond its own sales figures, influencing the broader Audi lineup and setting new standards for technology integration in sports cars.

As the automotive world began to shift towards electrification, the third-gen TT stood as a high watermark for the traditional sports car - a perfect blend of the analogue and digital worlds. Its reception and impact set the stage for the next evolution of the TT, leaving many to wonder how this beloved model would adapt to the electric age while maintaining the unique character that had made it a favorite among car enthusiasts for over two decades.

Chapter 10: Inside the TT: Interior Innovations and Comfort

Section 10.1: The Philosophy Behind TT's Interior Design

The Audi TT's interior design has always been a testament to the brand's commitment to innovation and driver-centric philosophy. From its inception, the TT's cabin has been more than just a place to sit; it's a carefully crafted environment that enhances the driving experience and embodies Audi's design ethos.

At the heart of the TT's interior design is a steadfast focus on the driver. The cockpit-style layout wraps around the driver, creating an intimate connection between the human and the machine. This approach is evident in every aspect of the interior, from the positioning of controls to the angle of the center console. It's a design that whispers to the driver, "This car was built for you."

Complementing this driver-centric approach is a commitment to minimalist aesthetics. The TT's interior embodies the 'less is more' principle, presenting a clean, uncluttered dashboard that has become a hallmark across generations. This minimalism isn't just about looks;

it's functional, reducing distractions and allowing the driver to focus on the road ahead.

However, minimalism doesn't mean compromise on luxury. Audi's use of premium materials and attention to build quality creates an environment that feels as luxurious as it is sporty. Fine Nappa leather adorns the seats, while brushed aluminum accents provide a technical, modern feel. Every surface, every stitch, every detail is crafted to meet the highest standards, reinforcing the TT's position as a premium sports car.

The TT's interior design also strikes a delicate balance between sportiness and everyday usability. While the low-slung seating position and driver-focused layout speak to its performance credentials, thoughtful ergonomics ensure comfort for daily driving. The perfectly positioned gear shifter falls naturally to hand, while supportive sports seats provide comfort even on longer journeys.

The most striking aspect of the TT's interior design philosophy is its seamless blend of form and function. Every element serves a purpose while contributing to the overall aesthetic. A prime example of this is the integration of air vents into the dashboard design. These circular vents, a signature TT feature, aren't just functional; they're an integral part of the interior's visual appeal.

This philosophy of functional aesthetics extends throughout the cabin. The flat-bottomed steering wheel isn't just a sporty touch; it provides better legroom and easier ingress and egress. The minimalist center console isn't just visually appealing; it creates a more spacious feel in the compact interior.

Over the years, this core philosophy has evolved with advancements in technology and changing design trends. However, the essence remains the same: a driver-focused, minimalist, premium interior that perfectly balances form and function.

The TT's interior design philosophy has not only defined the model but has also influenced Audi's broader design language and set benchmarks in the automotive industry. It proves that a car's interior can be as exciting and innovative as its exterior, and that a truly great sports car delivers its thrills not just in how it drives, but in how it feels from behind the wheel.

Section 10.2: First Generation (1998-2006): Setting the Stage

The first-generation Audi TT, introduced in 1998, set a new standard for sports-car interiors. Its cabin design was nothing short of trailblazing, blending form and function in a way rarely seen in the automotive world.

At the heart of the TT's interior design was a bold circular theme that permeated every aspect of the cabin. This was most prominently displayed in the circular air vents that became an instant TT signature. These vents weren't just functional; they were sculptural elements that added a sense of drama and purpose to the dashboard. The circular motif extended to other components as well, from the instrument gauges to the door handles, creating a cohesive and visually striking environment.

Aluminum accents played a crucial role in defining the first-generation TT's interior aesthetic. The extensive use of this material created a modern, technical feel that perfectly complemented the car's exterior design. Aluminum-ringed gauges and a center console trimmed in brushed aluminum gave the cabin a cool, industrial edge, setting it apart from its competitors. This use of metal wasn't just about looks; it also conveyed a sense of quality and durability that buyers in this segment demanded.

One of the most interior options for the first-generation TT was the baseball-stitched leather seats. This unique treatment showcased Audi's attention to detail and willingness to think outside the box. The thick, contrasting stitching on the seats and door panels was reminiscent of a baseball glove, adding a touch of playful nostalgia to the otherwise modern interior. It was details like these that helped cement the TT's reputation as a car that prioritized design and craftsmanship.

The layout of the first-gen TT's interior was unabashedly driver-focused. The center console was angled towards the driver, putting all controls within easy reach. This driver-centric approach extended to the placement of the gear shifter and the design of the steering wheel, both optimized for engagement and control.

For those who wanted to emphasize the TT's performance heritage, Audi offered a Quattro interior package. This option included special touches like a Quattro-branded shift knob and steering wheel, reminding occupants of the car's all-wheel-drive prowess. These elements not only added a sporty flair but also connected the cabin to Audi's rallying history.

The first-generation TT's interior was a bold statement of intent from Audi. It demonstrated that a sports car's cabin could be just as well-designed as its exterior. The combination of unique design elements, high-quality materials, and driver-focused ergonomics sets a new benchmark in the segment. More than just a place to sit while driving, the TT's interior became a key part of the car's appeal, offering a visually exciting environment that was a joy to interact with. This inaugural design laid the foundation for future generations of the TT, establishing design principles and signature elements that would evolve but never disappear in subsequent models.

Section 10.3: Second Generation (2006-2014): Refining the Concept

The second-generation Audi TT, introduced in 2006, marked a significant evolution in the car's interior design. Building upon the trailblazing concepts of its predecessor, this iteration refined and modernized the TT's cabin, creating a space that was both familiar to fans of the original and excitingly fresh for newcomers.

At the heart of this evolution was the circular theme that had become synonymous with the TT. Audi's designers took the signature circular air vents and gave them a more sophisticated, integrated look. The new vents featured a turbine-like design, adding a touch of mechanical elegance to the dashboard. This update maintained the TT's visual identity while firmly pushing it into the new millennium.

Ergonomics improved markedly in this generation. The seats were redesigned with better lateral support, catering to the TT's sporting pretensions while ensuring comfort during longer journeys. The driving position was fine-tuned, and controls were placed with even greater consideration for the driver's reach and visibility. These refinements enhanced the TT's reputation as a driver's car, making it more engaging and comfortable to operate.

This generation also saw the introduction of more advanced technology into the TT's interior. The optional Audi Magnetic Ride system, for instance, brought a new level of customization to the driving experience. Controls for this system were seamlessly integrated into the cockpit, allowing drivers to adjust the car's handling characteristics on the fly. This addition highlighted Audi's commitment to blending cutting-edge technology with user-friendly design.

Material quality took a significant leap forward in the second-generation TT. Audi pushed to make the interior feel more premium, using soft-touch materials on the dashboard and door panels. The leather upholstery option was improved, offering a more luxurious feel and appearance. These enhancements elevated the TT's interior ambiance, reinforcing its position as a premium sports car.

Customization options were expanded for this generation, allowing buyers to personalize their TT's interior to a greater degree. Audi offered an expanded range of interior color schemes and trim options, from classic black leather to more adventurous combinations. This move acknowledged the TT's status as a design-led vehicle and catered to owners who viewed their cars as expressions of personal style.

The overall design language of the second-generation TT interior struck a balance between the avant-garde approach of the original and a more mature, sophisticated aesthetic. The dashboard retained its driver-centric layout but featured a cleaner, more streamlined design. The center console was refined, with controls grouped more logically and finished with higher-quality switchgear.

Audi also improved the TT's practicality without compromising its sports car ethos. Storage solutions were cleverly integrated into the design, with more usable door pockets and center console space. The rear seats, while still compact, were made slightly more usable, adding to the car's versatility.

Lighting played a more prominent role in this generation's interior design. Ambient lighting was introduced, subtly highlighting key areas of the cabin and creating a more atmospheric environment during night driving. This feature, which would become increasingly important in later Audi models, made its debut in the TT.

In essence, the second-generation Audi TT's interior represented a thoughtful evolution of the original concept. It maintained the unique character that had made the first TT so iconic while addressing areas for improvement and incorporating new technologies. This approach ensured that the TT remained at the forefront of interior design in the sports car segment, setting the stage for even more innovative changes with the third generation.

Section 10.4: Third Generation (2014-present): Technological Revolution

The third-generation Audi TT, introduced in 2014, marked a significant leap forward in interior design and technology. This latest iteration of the legendary sports car showcased Audi's commitment to innovation and redefined the driver-centric cockpit.

At the heart of this technological revolution is the Audi Virtual Cockpit, a game-changing digital instrument cluster that has since become a hallmark of Audi's most advanced models. This 12.3-inch high-resolution display replaces not only the traditional analog gauges but also eliminates the need for a central infotainment screen. The result is a sleek, uncluttered dashboard that puts all essential information directly in the driver's line of sight. With customizable layouts and the ability to display everything from speed and RPM to navigation maps and media information, the Virtual Cockpit offers unprecedented versatility and user-friendliness.

Complementing the Virtual Cockpit is a newly designed multifunctional steering wheel. This ergonomic marvel allows drivers to control most of the car's functions without ever taking their hands off the wheel. From adjusting the audio system to scrolling through different Virtual Cockpit displays, the steering wheel becomes a command center for the entire vehicle. This integration of controls not only enhances safety by minimizing distractions but also reinforces the TT's driver-focused ethos.

The introduction of the Virtual Cockpit allowed Audi's designers to reimagine the center console. Gone is the traditional array of buttons and knobs, replaced by a minimalist design that emphasizes the TT's clean, uncluttered aesthetic. This decluttered approach extends to removing the central screen, with all information now displayed in the Virtual Cockpit. The result is a dashboard that's not only visually striking but also more intuitive to use, allowing drivers to focus on the joy of driving.

One of the most innovative features of the third-generation TT's interior is the integration of climate controls into the air vents. Each circular vent, a TT signature design element since the first generation, now houses a small digital display and a control dial in its center. This clever integration enables precise temperature adjustments and airflow control, while maintaining the dashboard's clean lines. It's a perfect example of how the TT's designers have seamlessly blended form and function.

The materials used in the latest TT interior continue Audi's tradition of premium quality while pushing boundaries. Alongside the fine leather and brushed aluminum that have long been TT staples, Audi now offers options like Alcantara and carbon fiber trim. These advanced materials not only enhance the car's sporty feel but also demonstrate Audi's commitment to using cutting-edge materials to create a truly special driving environment.

The third-generation TT's interior is a testament to Audi's forward-thinking approach to car design. By embracing digital technology and innovative design solutions, Audi has created a cockpit that feels more like a jet fighter than a traditional car. This high-tech environment, combined with the TT's signature styling cues and premium materials, results in an interior that's both futuristic and instantly recognizable as a TT.

Moreover, this generation of the TT has set new standards not just within the Audi brand, but across the automotive industry. The Virtual Cockpit technology, first introduced in the TT, has since proliferated across Audi's lineup and inspired similar systems from other manufacturers. Similarly, the minimalist, tech-forward approach to interior design has influenced sports cars and mainstream vehicles alike.

In essence, the third-generation Audi TT's interior represents a technological revolution wrapped in the familiar, much-loved TT design language. It proves that, even after decades in production, the TT remains a trendsetter, pushing the boundaries of what's possible in automotive interior design and technology.

Section 10.5: Comfort Features: Balancing Sport and Luxury

The Audi TT has always walked a fine line between being a pure sports car and a comfortable daily driver. This balance is most evident in its comfort features, which have evolved significantly over the years, creating an interior that's both sporty and luxurious.

One of the most crucial elements in achieving this balance is the seat design. The TT's seats have undergone a remarkable evolution since the first generation. While they've always been sporty, each iteration has improved both support and comfort. The latest generation showcases this perfectly with its S sport seats. These seats feature pneumatic side bolsters that can be adjusted to hug the driver and passenger snugly during spirited driving, while still providing comfort for longer journeys. The option for heated seats has also been a welcome addition for those in colder climates, proving that performance and comfort can indeed coexist.

Climate control is another area where the TT has made significant strides. The early models featured a basic yet effective system, but as the TT moved upmarket, so did its climate control capabilities. The introduction of the deluxe automatic air conditioning system in later models was a breakthrough. This system allows for precise temperature control and even accounts for the intensity when adjusting the cabin temperature. It's a small but significant feature that showcases Audi's attention to detail in creating a comfortable driving environment.

The TT's audio systems have also seen considerable upgrades over the years, catering to audiophiles who demand high-quality sound even in a sports car. While the base systems have always been competent, it's the optional upgrades that truly shine. The pinnacle of this is the Bang & Olufsen sound system available in the latest models. This system, with its 680 watts of power and 12 speakers, transforms the TT's cabin into a concert hall on wheels. It's a perfect example of how Audi has incorporated luxury features without compromising the car's sporty character.

Lighting has played an increasingly important role in the TT's interior comfort features. What started as basic illumination has evolved into a sophisticated LED interior lighting package. This customizable system not only provides practical illumination but also allows drivers to set the cabin's mood. From subtle ambient lighting to more dramatic color schemes, this feature adds an extra layer of personalization to the TT's interior.

One of the most challenging aspects of designing a sports car interior is incorporating practical storage solutions without disrupting the clean, driver-focused design. Audi has tackled this challenge admirably in the TT. Clever placement of cupholders ensures that they're accessible without being intrusive. The optional storage package adds compartments and nets that provide additional space for small items, proving that even a sports car can be practical.

All these comfort features come together to create an interior that's much more than the sum of its parts. The Audi TT offers the comfort and amenities of a luxury car while maintaining a focused, driver-centric feel. It's this careful balance that has helped the TT appeal to a wide range of buyers, from weekend racers to daily commuters.

The evolution of these comfort features in the TT also reflects broader trends in the automotive industry. As consumers have come to expect more from their vehicles, even dedicated sports cars like the TT have had to adapt, offering more comfort and technology alongside performance. The TT has managed this transition gracefully, never losing sight of its sporty roots while continually improving the everyday usability and comfort that modern drivers demand.

In essence, the Audi TT's comfort features are a masterclass in automotive interior design. They demonstrate that with careful thought and innovative solutions, it's possible to create a sports car interior that's both thrilling and comfortable, ready for both the racetrack and the daily commute.

Section 10.6: Special Editions and Performance Variants

The Audi TT's interior has always been a canvas for innovation and creativity, and nowhere is this more evident than in its special editions and performance variants. These unique models showcase Audi's commitment to pushing boundaries and catering to the most discerning enthusiasts.

At the pinnacle of the TT range sits the TT RS, a high-performance variant that takes the already driver-focused interior to new heights. The TT RS features a steering wheel, instantly recognizable by its Alcantara trim and the eye-catching red RS badge at the 12 o'clock position. This wheel isn't just for show; its flat-bottom design and enlarged paddle shifters provide enhanced control and feedback during spirited driving.

The RS-specific sport seats are another highlight, offering increased lateral support and an integrated headrest design that echoes the seats found in professional racing cars. These seats are often upholstered in fine Nappa leather with a diamond-pattern stitching, further emphasizing the model's premium positioning.

Anniversary editions of the TT have provided Audi designers with the opportunity to create truly special interiors that celebrate the model's heritage. The 20th Anniversary Edition, for example, featured a unique color scheme that paid homage to the original TT concept car. This included Moccasin Brown leather seats with yellow contrast stitching, a combination that instantly transports drivers back to the TT's debut. Special badging and embossed logos on the seats serve as subtle reminders of the car's significance.

For those seeking the ultimate in personalization, Audi offers the Exclusive program for the TT. This bespoke service allows customers to specify unique leather colors, stitching patterns, and trim materials that aren't available on standard models. Want your TT's interior to match your favorite watch or pair of shoes? With Audi Exclusive, it's possible. This program has resulted in some truly one-of-a-kind TT interiors, from subtle and sophisticated to bold and avant-garde.

Performance enthusiasts who opt for the Competition package are treated to an interior that emphasizes the TT's sporting credentials. The package typically includes an Alcantara-wrapped steering wheel, a material known for its excellent grip. Carbon fiber trim replaces standard aluminum accents, reducing weight while adding a motorsport-inspired aesthetic. The seats in Competition package cars often feature contrast stitching in bold colors like red or yellow, adding a visual pop to the cabin.

Audi has also collaborated with other luxury brands to create limited edition TTs with truly unique interiors. A standout example is the Ultra LE, designed in partnership with high-end luggage manufacturer Rimowa. This special edition featured an interior trimmed in Rimowa's signature grooved aluminum, creating a striking visual link between the car and the company's renowned suitcases. The result was an interior unlike anything else on the road, showcasing the potential for cross-industry collaboration in automotive design.

These special editions and performance variants demonstrate the TT's interior design's versatility and enduring appeal. Whether it's through high-performance enhancements, anniversary celebrations, bespoke customizations, or unexpected collaborations, Audi continues to find new ways to make the TT's cabin memorable. Each of these unique interiors tells a story, adding another chapter to the rich history of this celebrated sports car. They serve not only to attract collectors and enthusiasts but also to inspire future developments in the TT range and beyond.

Section 10.7: The TT's Interior Influence on the Automotive Industry

The Audi TT's interior design has left an indelible mark on the automotive industry, influencing not only other sports cars but also setting new standards for interior design across various vehicle segments. From its inception, the TT has been a trendsetter, pushing boundaries and challenging conventional wisdom about what a sports car's interior should look and feel like.

One of the most significant ways the TT has influenced the industry is by setting trends in sports car interiors. The clean, minimalist aesthetic that has been a hallmark of the TT since its first generation has inspired many competitors to rethink their approach to interior design. The driver-focused layout, with its cockpit-like feel, has become increasingly common in sports cars across different brands. This approach not only enhances the driving experience but also creates a sense of connection between the driver and the vehicle, a connection that many manufacturers now strive to emulate.

The TT has also been at the forefront of pushing technological boundaries in the automotive interior. The most notable example of this is the Audi Virtual Cockpit, which debuted in the third-generation TT before being adopted across Audi's broader model range and influencing digital instrument clusters throughout the industry.

This innovative system not only streamlined the interior design by eliminating the need for a central infotainment screen but also enhanced the driver's interaction with the vehicle's systems. The TT's role as a testbed for Audi's interior innovations has allowed the brand to perfect new technologies in a premium sports car setting before rolling them out to other models.

Moreover, the TT has played a crucial role in redefining expectations within its segment. Before the TT's arrival, many sports cars prioritized performance over interior quality and comfort. The TT demonstrated that a sports car could offer a premium, well-crafted interior without compromising its sporting credentials. This shift has raised the bar for interior quality across the sports car segment, with competitors now expected to offer a level of fit, finish, and material quality that matches the TT's standards.

The TT's success in balancing form and function has also had a widespread impact. Its interior design proves that a sports car can be both beautiful and practical, a concept that many manufacturers have since embraced. The clever integration of features like the climate controls within the air vents in the latest generation TT showcases how functional elements can be seamlessly incorporated into the overall design aesthetic. This approach to functional aesthetics has inspired other automakers to think creatively about how to blend necessary features with appealing design.

Looking to the future, the current TT interior is likely to influence upcoming Audi models and industry designs more broadly. We can expect to see even more digital integration in future sports cars, with customizable interfaces becoming increasingly common.

The trend towards minimalist, uncluttered interiors is likely to continue, with physical buttons being replaced by touch-sensitive surfaces or voice-activated controls. Additionally, the concept of a fully customizable interior space, where drivers can personalize everything from the layout of their digital displays to the ambient lighting, may become more prevalent, building on the foundations laid by the TT.

In conclusion, the Audi TT's interior has been more than just a comfortable place to sit while driving a sports car. It has been a showcase of innovation, a quality standard-setter, and a glimpse into the future of automotive interior design. As the automotive industry continues to evolve, particularly with the advent of electric and autonomous vehicles, the principles of driver-focused design, technological integration, and balanced design exemplified by the TT are likely to remain influential, shaping how we interact with our vehicles for years to come.

The Audi TT: A Journey Through Design and Innovation

Chapter 11: The TT's Cultural Impact

Section 11.1: The TT's Initial Reception in the Automotive World

When the Audi TT debuted in 1998, it sent shockwaves through the automotive industry. Car enthusiasts and the automotive press alike were instantly captivated by its game-changing design and impressive performance capabilities. The TT's unique blend of retro-inspired aesthetics and modern engineering struck a chord with critics and consumers, earning it immediate acclaim.

Automotive journalists praised the TT for its bold, avant-garde styling that set it apart from anything else on the road. The car's rounded silhouette, minimalist interior, and innovative use of aluminum in its construction were frequently cited as trailblazing features. Many reviewers noted that the TT felt like a concept car that had somehow made it to production, preserving the purity of its original design vision.

The TT's impact was further solidified by the numerous awards and accolades it received upon release. It was named "Car of the Year" by several prominent automotive publications, and its design team, led by Freeman Thomas and Peter Schreyer, received widespread recognition for their innovative work. The car's success at award ceremonies helped cement its status as a trailblazer in the automotive world.

Comparisons to contemporary sports cars inevitably followed, with the TT often coming out on top for style and uniqueness. While some critics argued that more traditional sports cars offered superior performance, most agreed that the TT's combination of style, handling, and everyday usability made it a compelling package. The car was frequently pitted against rivals like the Porsche Boxster and BMW Z3, often earning praise for its character and design-forward approach.

The TT's influence on automotive design trends was almost immediate. Its clean lines, minimalist aesthetic, and innovative use of materials inspired numerous imitators across the industry. The car's impact could be seen in everything from other sports cars to more mundane sedans and hatchbacks, as manufacturers sought to capture some of the TT's design magic.

The most telling sign of the TT's initial impact was the rapid formation of TT owners' clubs and enthusiast groups. Early adopters of the car were quick to form communities, both online and in person, dedicated to celebrating and modifying their TTs. These groups played a crucial role in building the car's reputation and fostering a strong enthusiast culture around the model.

The TT's initial reception in the automotive world was nothing short of phenomenal. It captured the imagination of car enthusiasts, design aficionados, and the general public alike. This strong start laid the foundation for the TT's enduring appeal and its eventual transition from mere automobile to cultural icon. The ripples from the TT's splash onto the automotive scene would continue to be felt for years to come, influencing car design, enthusiast culture, and Audi's brand identity in profound ways.

Section 11.2: The TT in Motorsports and Its Impact on Enthusiast Culture

The Audi TT's journey from showroom to racetrack marked a significant chapter in its evolution, cementing its status as more than just a stylish sports car. From the moment it hit the circuits, the TT proved it had the performance to match its looks, captivating motorsports enthusiasts and elevating its reputation among car aficionados.

The TT's racing pedigree began to take shape shortly after its debut, with appearances in various racing series around the world. One of the most notable was its participation in the German VLN Endurance Racing Championship at the legendary Nürburgring. Here, the TT demonstrated its ability to compete with purpose-built race cars, often punching above its weight. The car's compact size, balanced chassis, and Quattro all-wheel-drive system proved to be formidable assets on the challenging Nordschleife circuit.

In addition to endurance racing, the TT found success in touring car championships. The British Touring Car Championship (BTCC) saw specially prepared TTs battling against rivals from other manufacturers, often securing podium finishes and occasionally clinching victories. These appearances not only demonstrated the TT's racing prowess but also provided valuable data for Audi's engineers to refine the road-going versions.

One of the most impressive feats in the TT's motorsport history was its participation in the Pikes Peak International Hill Climb. The grueling uphill race, known for its treacherous turns and rapidly changing weather, saw modified TTs tackle the mountain with aplomb. The car's all-wheel-drive system and turbocharged engines were particularly well-suited to the challenging ascent, resulting in several class victories and impressive overall finishes.

These racing successes had a profound impact on the enthusiast culture surrounding the TT. Owners' clubs began organizing track days, allowing enthusiasts to experience their TTs in a controlled racing environment. The car's performance on the track inspired many owners to modify their vehicles, fueling a vibrant aftermarket scene. Tuning companies developed performance upgrades specifically for the TT, ranging from simple ECU remaps to complete engine overhauls, as well as aerodynamic kits inspired by the racing versions.

The TT's racing achievements also influenced Audi's approach to its performance car lineup. The lessons learned on the track were often incorporated into special edition models, such as the TT RS, which brought near-racecar levels of performance to the road. This racing DNA became a key selling point for Audi, attracting enthusiasts who wanted a taste of motorsport in their daily driver.

Enthusiast events centered around the TT began to proliferate, with owners gathering to share their passion for the car. These ranged from local meets to large-scale international events, often featuring both stock and heavily modified TTs. Some events even incorporated amateur racing or time attack competitions, further blurring the line between the TT's road and racing personas.

The TT's influence extended beyond just Audi enthusiasts. Its success in various motorsport disciplines earned respect from the broader performance car community. The car became a common sight at track days and autocross events, often competing against more expensive and exotic machinery.

As the TT evolved through its generations, its motorsport heritage remained a crucial part of its identity. Each new iteration was eagerly anticipated by enthusiasts, who looked forward to seeing how Audi would incorporate the latest racing technology and lessons into the road car.

The TT's journey in motorsports and its impact on enthusiast culture demonstrate how a well-designed sports car can transcend its original purpose. From a stylish road car to a competitive racer and a cornerstone of car culture, the Audi TT proved that performance and passion go hand in hand. Its racing successes not only validated its capabilities but also fostered a community of enthusiasts who continue to celebrate and push the boundaries of this legendary sports car.

Section 11.3: The Audi TT in Popular Media

The Audi TT's sleek design and cultural significance have made it a favorite in popular media, appearing in various forms of entertainment and capturing the public's imagination far beyond the automotive world.

In major films and television shows, the Audi TT has often been featured as a symbol of style and sophistication. Its appearances range from high-octane action sequences to subtle background placements that reinforce a character's taste and status. For instance, in the 2004 sci-fi thriller "I, Robot," the Audi TT concept car made a memorable appearance, showcasing a futuristic version of the already forward-thinking design. This cameo not only highlighted the car's aesthetic appeal but also solidified its association with cutting-edge technology and innovation.

The gaming industry has also embraced the Audi TT, featuring it prominently in numerous racing games and simulators. From popular franchises like "Forza Motorsport" and "Gran Turismo" to more casual mobile racing games, the TT has been a staple choice for virtual drivers. These digital representations have allowed millions of gamers worldwide to experience the thrill of driving a TT, further cementing its place in popular culture and potentially influencing real-world purchasing decisions.

Celebrity ownership and endorsements have played a significant role in elevating the TT's status. Notable figures from various fields, including actors, musicians, and athletes, have been spotted driving the Audi TT, which has often led to increased media coverage and public interest. These high-profile owners serve as unofficial brand ambassadors, associating the car with success, luxury, and good taste.

The Audi TT: A Journey Through Design and Innovation

In the realm of automotive journalism, the Audi TT has been a frequent subject of reviews, comparison tests, and feature articles. Its unique design and driving characteristics have made it a favorite among motoring journalists, who often praise its blend of style and performance. The car's regular appearances in prestigious automotive magazines and websites have kept it in the public eye and contributed to its enduring appeal.

The advent of social media has opened up new avenues for the Audi TT to maintain its cultural relevance. Enthusiast groups and fan pages dedicated to the TT have flourished on platforms like Facebook, Instagram, and Reddit. These online communities share photos, modification ideas, and personal stories, creating a vibrant digital ecosystem around the car. User-generated content featuring the TT, from artistic photographs to restoration projects, has helped maintain interest in the model across generations of car enthusiasts.

The Audi TT's presence in popular media extends beyond traditional channels. It has been featured in music videos, serving as a visual shorthand for luxury and style. Artists across genres have incorporated the TT into their visual aesthetics, associating their brand with the car's trendsetting design.

Moreover, the TT has inspired various forms of merchandise and collectibles, from scale models to art prints, allowing fans to celebrate their appreciation for the car beyond ownership. This merchandising has further integrated the TT into everyday life, making it a recognizable icon even among non-car enthusiasts.

The Audi TT's journey through popular media demonstrates its unique position as more than just a car. It has become a cultural touchstone, representing a perfect blend of design, performance, and aspiration. Its consistent presence across various media platforms has ensured that the TT remains relevant and desirable, even as automotive trends evolve. This media exposure has not only bolstered the TT's reputation but has also contributed significantly to

Audi's overall brand image, positioning the company as a producer of stylish, innovative vehicles that capture the public's imagination.

As we continue to consume media in increasingly diverse ways, the Audi TT's role as a pop culture symbol seems set to endure, inspiring new generations of car enthusiasts and design aficionados alike.

Section 11.4: The TT as a Design Icon

The Audi TT's impact extends far beyond the automotive world, cementing its status as a true design icon. From its inception, the TT's bold, innovative design caught the attention of creative professionals across industries, inspiring a wave of admiration and imitation that continues to this day.

In design circles, the TT is often held up as a masterclass in form and function. Its clean lines, perfect proportions, and innovative use of materials have earned it recognition from prestigious design institutions worldwide. The car's inclusion in numerous design exhibitions and museums stands as a testament to its artistic merit. For instance, the Museum of Modern Art in New York featured the TT in its "Different Roads: Automobiles for the Next Century" exhibit, highlighting its significance in the evolution of automotive design.

The TT's influence has rippled through industrial and product design, inspiring everything from furniture to consumer electronics. Designers have drawn inspiration from its silhouette and minimalist aesthetic, applying these principles to create sleek, modern products that echo the TT's timeless appeal. The car's impact is evident in the curved edges of smartphones, the streamlined shapes of modern appliances, and the clean lines of contemporary furniture.

Recognizing the TT's broader design appeal, Audi has collaborated with designers from other fields, further solidifying the car's status as a cross-disciplinary icon. These partnerships have resulted in limited-edition models and design objects that showcase TT's versatility as a source of inspiration. For example, a collaboration with renowned fashion designer Marc Newson led to a concept car that pushed the boundaries of automotive design even further, blending the TT's shape with futuristic elements.

Perhaps most significantly, the TT's design language has permeated Audi's broader lineup, influencing the aesthetic direction of the entire brand. The bold simplicity and attention to detail that defined the original TT can now be seen across Audi's range, from compact hatchbacks to luxury sedans. This design philosophy, characterized by clean surfaces, precise lines, and a focus on high-quality materials, has become a hallmark of the Audi brand, helping it stand out in a crowded automotive market.

The TT's impact on design extends to its interior as well. Its driver-focused cockpit, with its innovative use of circular themes and high-quality materials, set new standards for sports car interiors. This approach has influenced not only other Audi models but also competitors' designs, raising the bar for interior quality and ergonomics across the industry.

As a design icon, the Audi TT has achieved something rare in the automotive world: it has transcended its original purpose to become a symbol of good design. Its influence can be felt not just on roads around the world, but in design studios, art galleries, and even in everyday objects. The TT serves as a reminder of the power of bold, innovative design to capture imaginations and shape culture, solidifying its place not just in automotive history but in the broader pantheon of design excellence.

Section 11.5: The TT's Impact on Audi's Brand Image

The Audi TT's influence extends far beyond its role as a sports car, playing a pivotal part in shaping and elevating Audi's overall brand image. When the TT first debuted, it marked a turning point for Audi, propelling the company into the upper echelons of premium automotive brands.

The Audi TT: A Journey Through Design and Innovation

The TT's sleek, avant-garde design and impressive performance capabilities significantly enhanced Audi's perception as a premium brand. It showcased the company's ability to produce not just reliable luxury vehicles, but also cutting-edge sports cars that could compete with the best in the world. The TT's unique blend of style and substance resonated with consumers and critics alike, positioning Audi as a forward-thinking, design-conscious automaker.

In Audi's marketing and advertising campaigns, the TT often took center stage. Its silhouette and eye-catching design made it the perfect ambassador for the brand's "Vorsprung durch Technik" (Progress through Technology) slogan. The TT featured prominently in print ads, television commercials, and digital marketing efforts, helping to create a strong visual association between Audi and innovative design. These campaigns not only promoted the TT itself but also reinforced Audi's image as a manufacturer of sophisticated, technologically advanced vehicles.

The TT's influence extended beyond its own model line, shaping Audi's design philosophy for other vehicles in its lineup. Elements of the TT's design language, such as its clean lines, bold grille, and driver-focused interior, began to appear across Audi's range. This cohesive design approach helped to create a strong, recognizable brand identity for Audi, setting it apart from its competitors in the luxury car market.

The TT's success had a tangible impact on Audi's sales and market position. As the model gained popularity and critical acclaim, it attracted new customers to the brand, many of whom went on to purchase other Audi vehicles. The halo effect of the TT helped boost sales across Audi's lineup, contributing to the company's growing market share in the premium automotive segment.

The accolades and recognition garnered by the TT also reflected positively on Audi as a whole. Awards for design excellence, performance, and innovation not only celebrated the TT but also reinforced Audi's reputation as a manufacturer of high-quality, desirable vehicles. These accolades provided valuable marketing opportunities and helped to build consumer trust in the Audi brand.

Perhaps most importantly, the TT played a crucial role in differentiating Audi from its German luxury car rivals. While BMW was known for its sporting sedans and Mercedes-Benz for its luxurious comfort, the TT helped establish Audi as the design-led, technologically innovative choice. This unique positioning has continued to serve Audi well, allowing it to carve out a distinct identity in a crowded marketplace.

The TT's impact on Audi's brand image cannot be overstated. It transformed perceptions of the company, elevating it from a maker of solid, respectable cars to an authentic luxury brand with a flair for design and innovation. The TT's legacy continues to influence Audi's image today, serving as a reminder of the company's commitment to pushing boundaries and challenging conventions in automotive design and engineering.

Section 11.6: The TT in Art and Popular Culture

The Audi TT's influence extends far beyond the automotive world, permeating various facets of art and popular culture. Its design and cultural significance have made it a muse for artists, writers, and creatives across different media.

In the realm of visual arts, the TT has inspired numerous artistic interpretations. Renowned automotive artists have captured its sleek lines and silhouette in paintings, sculptures, and digital artworks. These pieces often highlight the car's unique design elements, such as its rounded roofline and bold wheel arches, elevating the TT from a mere vehicle to a subject of artistic expression. Galleries and exhibitions dedicated to automotive art frequently feature the TT, cementing its status as a design legend worthy of creative celebration.

Literature and automotive writing have also embraced the Audi TT as a subject of interest. Numerous books and articles have been penned about the car's design philosophy, cultural impact, and driving experience. Automotive journalists and enthusiasts have waxed poetic about the TT's handling, performance, and aesthetic appeal, often using it as a benchmark for discussing modern sports car design. The TT has even made appearances in fiction, sometimes serving as a character's prized possession or a symbol of aspiration and success.

The music industry has not been immune to the TT's charms either. The car has made notable appearances in music videos, often associated with themes of luxury, style, and sophistication. Its sleek form has graced album covers, particularly in genres that emphasize cutting-edge design and urban cool. The TT's presence in these visual media has helped reinforce its image as a symbol of contemporary style and automotive excellence.

Fashion and lifestyle brands have also recognized the TT's appeal, leading to various collaborations that extend the car's influence beyond the garage. High-end fashion houses have drawn inspiration from the TT's design language, incorporating its curves and lines into clothing and accessory collections. Limited edition TT-inspired products, from watches to luggage sets, have hit the market, allowing enthusiasts to embrace the TT lifestyle even when they're not behind the wheel.

Perhaps most significantly, the Audi TT has become a symbol of modern design in popular culture. Its shape is instantly recognizable, often used in media as shorthand for cutting-edge style and German engineering prowess. The TT's appearance in a film or TV show immediately establishes a character or setting as design-conscious and contemporary. This cultural cachet has transcended the automotive sphere, making the TT a reference point in discussions about influential industrial design across all sectors.

The TT's integration into popular culture has also manifested in unexpected ways. It has been featured in art installations exploring the intersection of technology and design, used as a canvas for avant-garde artists pushing the boundaries of automotive aesthetics, and even inspired architectural elements in modern buildings. These diverse interpretations underscore the TT's versatility as a cultural touchstone, capable of sparking creativity across various disciplines.

As social media has become an increasingly dominant force in shaping popular culture, the Audi TT has found a new platform to showcase its enduring appeal. Enthusiasts share stunning photographs of their TTs in picturesque locations, custom modifications, and nostalgic throwbacks to earlier models. This digital celebration of the TT has introduced it to new generations, ensuring its cultural relevance continues well into the 21st century.

The Audi TT's journey from a pioneering sports car to a pop culture icon is a testament to the power of exceptional design. Its presence in art, literature, music, fashion, and digital media demonstrates how a well-conceived automobile can transcend its original purpose to become a symbol of an era's aesthetic sensibilities. As the TT continues to evolve, its cultural impact serves as a reminder of the profound influence that innovative design can have on society.

Section 11.7: The TT's Legacy and Lasting Impact

The Audi TT's journey from a trailblazing concept to a beloved sports car has left an indelible mark on the automotive landscape. Its legacy extends far beyond its sales figures or performance statistics, influencing the industry and popular culture in ways that continue to resonate today.

The TT's influence on subsequent sports car designs is undeniable. Its bold, minimalist aesthetic and iconic silhouette inspired a generation of designers to push boundaries and challenge conventional wisdom. Many modern sports cars, both from Audi and other manufacturers, bear subtle nods to the TT's trailblazing design language. The emphasis on clean lines, purposeful simplicity, and the perfect balance between form and function can be traced back to the TT's innovative approach.

The Audi TT: A Journey Through Design and Innovation

As the years have passed, early TT models have begun to appreciate, becoming sought-after collectibles. Enthusiasts and collectors alike recognize the historical significance of these early examples, particularly the first-generation models that most closely resemble the original concept car. This growing collectibility is a testament to the TT's enduring appeal and its essential place in automotive history.

In discussions about automotive design evolution, the TT is frequently cited as a pivotal moment. It represents a shift away from the angular, aggressive designs of the 1980s and early 1990s towards a more refined, sophisticated aesthetic. Design schools and automotive historians often use the TT as a case study in how a single model can redefine an entire segment and influence broader design trends.

The TT's success has had a profound impact on Audi's approach to future sports car development. It demonstrated that a bold, design-led approach could yield both critical acclaim and commercial success. This lesson has informed Audi's strategy for subsequent models, encouraging the brand to take calculated risks and push the envelope in terms of both aesthetics and engineering.

Perhaps most significantly, the TT has secured its place in automotive history as more than just a sports car. It stands as a symbol of turn-of-the-millennium optimism, a physical embodiment of the excitement and possibility that characterized the late 1990s and early 2000s. Its clean, futuristic design captured the spirit of a new era, making it as much a cultural artifact as a mode of transportation.

The TT's enduring appeal lies in its ability to transcend its mere status as an automobile. It has become a design icon, a pop culture reference point, and a benchmark against which other sports cars are measured. Even as newer models have been introduced, the original TT's influence continues to be felt, its DNA evident in Audi's current lineup and beyond.

The Audi TT: A Journey Through Design and Innovation

As we look to the future, the TT's legacy serves as a reminder of the power of visionary design. It stands as proof that a car can be more than the sum of its parts; it can be a work of art, a cultural touchstone, and a source of inspiration for generations to come. The Audi TT's journey from concept to icon is a testament to the enduring power of bold ideas and flawless execution in the automotive world.

Chapter 12: Looking Ahead: The Future of the Audi TT in the Electric Era

Section 12.1: The Electric Revolution and the Audi TT

The automotive industry is undergoing a seismic shift, with electric vehicles (EVs) rapidly moving from the fringes to the mainstream. This global transition towards electrification is reshaping the landscape for all car manufacturers, including Audi, and poses both challenges and opportunities for iconic models like the TT.

A combination of factors drives the push towards EVs. Environmental concerns, particularly about climate change and air pollution, have led to stricter emissions regulations worldwide. Many countries have announced ambitious plans to phase out the sale of new gasoline-powered cars. For instance, the United Kingdom and several European nations have set targets to ban the sale of new internal combustion engine vehicles by 2030 or 2035.

This regulatory pressure, combined with growing consumer awareness and demand for more sustainable transportation options, has accelerated the development and adoption of electric vehicles.

Audi, recognizing this shift, has made a bold commitment to electrification. The German automaker has pledged to launch only all-electric models from 2026 onwards, signaling a dramatic transformation of its product lineup. This decision reflects Audi's determination to position itself at the forefront of the electric revolution and maintain its status as a premium brand in the evolving automotive landscape.

However, electrifying a sports car like the Audi TT presents unique challenges. Sports cars are known for their lightweight design, agile handling, and emotive driving experience – characteristics that can be compromised by the heavy battery packs required for EVs. The sound of a high-performance engine, an integral part of the sports car experience, is absent in electric vehicles. Moreover, the TT's compact size may make it difficult to package a battery large enough to deliver both impressive performance and an acceptable range.

Yet electric powertrains also offer exciting opportunities to enhance sports car performance. Electric motors deliver instant torque, which could significantly improve the TT's already impressive acceleration. The potential for precise torque vectoring with multiple electric motors could revolutionize the car's handling characteristics, surpassing the capabilities of traditional all-wheel-drive systems.

In Audi's electric future, the TT could play a crucial role. Just as the original TT made a bold design statement and helped establish Audi's reputation for progressive aesthetics, an electric TT could serve as a technology flagship for the brand. It could showcase Audi's ability to combine cutting-edge electric performance with emotive design and driver engagement.

The electric TT might take on a role similar to that of the R8 e-tron, which demonstrated Audi's early electric vehicle technology in a high-performance package.

As Audi navigates this transition, the company must balance preserving the TT's essence with embracing the possibilities of electric propulsion. The challenge lies in creating an electric sports car that not only meets the performance expectations of TT enthusiasts but also appeals to a new generation of drivers who prioritize sustainability alongside driving excitement.

The electrification of the Audi TT is not just about changing its powertrain; it represents a reimagining of what a sports car can be in the 21st century. As we look ahead, the electric TT has the potential to blend the model's rich heritage with forward-thinking technology, setting a new benchmark for electric sports cars and redefining the concept of performance for the electric age.

Section 12.2: Potential Design Evolution

As the Audi TT enters the electric era, its design will undoubtedly evolve to meet new technological demands while maintaining its iconic status. The challenge for Audi's designers will be to preserve the essence of the TT's beloved silhouette while adapting it to the requirements of an electric vehicle.

One of the most crucial aspects of this evolution will be retaining the TT's signature rounded roofline and overall profile. This shape has been a hallmark of the TT since its inception, and it's likely to remain a key element in its electric future. However, designers may need to make subtle aerodynamic adjustments to maximize the range of electric vehicles. We might see a slightly more swept-back windshield or a more tapered rear end, but these changes would likely be implemented in a way that honors the TT's classic form.

The front-end design of the electric TT is where we're likely to see the most significant changes. Traditional internal combustion engines require large cooling grilles, but electric powertrains have different cooling needs. This opens up new possibilities for the TT's face. Designers could create a sleeker, more futuristic front end that pays homage to the original TT while clearly signaling its electric powertrain. We may see a sealed-off grille area with intricate lighting designs or a new interpretation of Audi's signature Singleframe grille, reimagined for the electric age.

The integration of EV-specific design elements will be crucial in distinguishing the electric TT from its predecessors. This could include incorporating active aerodynamic features, such as a retractable rear spoiler that deploys at high speeds to reduce drag and improve efficiency. The side profile might feature more pronounced sculpting to channel air efficiently around the vehicle. Designers might also incorporate subtle visual cues to indicate the car's electric nature, such as unique wheel designs optimized for aerodynamics or badging.

Inside the car, the TT's interior is likely to undergo a significant evolution to accommodate EV technology and meet the expectations of tech-savvy buyers. The center console could be reimagined to incorporate a larger touchscreen interface for managing the car's electric systems, range information, and advanced infotainment features. However, Audi designers would likely strive to maintain the driver-focused cockpit feel that has been a hallmark of the TT. We might see a blend of physical controls and digital interfaces, balancing the need for advanced functionality with the tactile experience that sports car enthusiasts appreciate.

Lighting will undoubtedly play a key role in the design of the electric TT. LED matrix headlights and OLED taillights could become even more intricate and expressive, serving as both functional elements and key styling features. These advanced lighting systems could offer new ways to accentuate the car's lines and create a visual signature, especially at night. Dynamic lighting effects could also be used to communicate the car's charging status or greet the driver upon approach, adding a new dimension of interactivity to the TT's design.

The use of materials in the electric TT's design will also likely evolve. To offset the weight of the battery pack, we might see more extensive use of lightweight materials, such as carbon fiber or advanced composites, in the body panels. The interior could feature a mix of sustainable materials and high-tech fabrics, reflecting both the car's performance heritage and its eco-friendly powertrain.

Color and finish options for the electric TT could also push boundaries. Audi might offer new, bold color choices that emphasize the car's futuristic nature, including options for color-shifting paints or unique matte finishes. Special-edition models could feature design elements that celebrate the TT's electric transformation.

In essence, the design evolution of the electric Audi TT will be a delicate balance of honoring its rich heritage while boldly stepping into the future. It will need to be immediately recognizable as a TT, while clearly communicating its advanced electric powertrain. This evolution represents not just a change for the TT, but a broader shift in sports car design for the electric age. If successful, the electric TT's design could once again set new standards in the automotive world, just as the original TT did over two decades ago.

Section 12.3: Performance and Driving Dynamics

The transition to electric power presents both challenges and opportunities for the Audi TT's performance and driving dynamics. As we look to the future, it's clear that Audi's engineers will need to harness the unique characteristics of electric powertrains to maintain and even enhance the TT's reputation for exhilarating performance.

At the heart of the electric TT is a cutting-edge powertrain configuration. A dual-motor setup could provide all-wheel drive capability, echoing the current TT's quattro system. This configuration would not only maintain the car's traction and stability but could also offer even more precise torque vectoring. By independently controlling the power sent to each wheel, the electric TT could achieve a level of handling finesse that surpasses its internal combustion predecessors.

Battery technology will play a crucial role in the electric TT's performance. The challenge lies in providing sufficient range without compromising the car's weight distribution and handling characteristics. Solid-state batteries, with their higher energy density, could be the key to this balancing act. These advanced batteries could provide the range needed for extended driving pleasure while maintaining the TT's ideal weight distribution. Moreover, strategically placing the battery pack could lower the car's center of gravity, further enhancing its cornering performance.

Handling has always been a hallmark of the TT, and the electric version will need to live up to this legacy. Audi's engineers will likely focus on developing an advanced chassis that can cope with the unique characteristics of an electric powertrain. This could include adaptive suspension systems that adjust in real time to different driving conditions and driver inputs. Advanced torque vectoring systems, made possible by the precise control of electric motors, could provide even more nuanced control over the car's dynamics, allowing drivers to carve through corners with unprecedented precision.

One of the most significant challenges in creating an electric sports car is addressing the lack of engine noise. The sound of a high-performance engine has long been an integral part of the sports car experience. For the electric TT, Audi sound engineers might develop a unique 'electric sound' that enhances the driving experience without mimicking a combustion engine.

This could involve a combination of carefully crafted external sounds for pedestrian safety and internal audio cues that provide feedback on vehicle speed and performance, creating an immersive auditory experience that's distinctly "TT" while embracing its electric nature.

In terms of performance targets, Audi is likely to aim high to ensure the electric TT can compete with other high-performance EVs. A 0-60 mph time under 3 seconds could be achievable thanks to the instant torque of electric motors. Top speed might be electronically limited to around 155 mph, balancing performance, efficiency, and range. However, the actual performance benchmark for the electric TT will likely be its overall driving experience – the way it responds to driver inputs, how it handles on twisting roads, and the confidence it inspires in its pilot.

The electric TT could also introduce new performance features unique to EVs. For instance, a regenerative braking system could be tuned to provide different levels of deceleration, allowing for one-pedal driving in certain situations. This could add a new dimension to the driving experience, particularly in urban environments or on winding mountain roads.

Thermal management will be another crucial aspect of the electric TT's performance. High-performance driving generates significant heat, and managing it is essential to maintaining consistent performance. Audi may develop advanced cooling systems for the batteries and motors, ensuring the car delivers peak performance lap after lap on a racetrack.

Ultimately, the goal for the electric TT will be to deliver a driving experience that's not just as good as its internal combustion predecessors, but in many ways superior. The instant torque, low center of gravity, and precise control offered by an electric powertrain could make the electric TT the best-performing version yet.

It will need to offer the same level of driver engagement and emotional connection that has made the TT a favorite among enthusiasts, while also showcasing the performance potential of electric technology.

The electric TT has the potential to redefine what a sports car can be in the electric age. Combining Audi's performance heritage with cutting-edge EV technology could offer a driving experience that's both familiar to TT fans and revolutionary in its capabilities. As we look to the future, it's clear that the electric TT has the potential to not just live up to its predecessor's legacy but to create a thrilling new chapter in the story of the sports car.

Section 12.4: Technology Integration

As the Audi TT enters the electric era, it's poised to become a showcase of cutting-edge automotive technology. The integration of advanced systems will not only enhance the driving experience but also redefine what it means to be a modern sports car.

At the forefront of this technological revolution are advanced driver assistance systems. While the TT has always been a driver's car, the electric version could feature a track mode that provides coaching to help achieve optimal lap times, similar to Porsche's Track Precision app. This system could analyze driving data in real-time, offering suggestions for improved braking points, racing lines, and acceleration out of corners. It's a feature that would appeal to enthusiasts looking to hone their skills without compromising the pure driving experience that the TT is known for.

Connectivity and infotainment systems are set to take a quantum leap forward in the electric TT. An augmented reality heads-up display could provide navigation and performance data without distracting from the road. Imagine racing line projections overlaid on the track ahead or real-time energy consumption data floating in your peripheral vision. This technology would not only enhance the driving experience but also improve safety by keeping the driver's eyes on the road.

The electric TT is likely to embrace over-the-air updates and customization. This software-first approach could allow owners to purchase temporary performance upgrades, such as increased power output for track days. Audi could offer a range of digital performance packages, allowing drivers to tailor their TTs' characteristics to suit different driving scenarios. This flexibility adds a new dimension to the ownership experience, providing multiple cars in one.

Integration with smart home and grid technology is another area where the electric TT could shine. The car could communicate with home energy systems to optimize charging times and reduce electricity costs. Imagine your TT automatically charging when electricity rates are lowest or using its battery to power your home during peak hours. This level of integration would position the TT not just as a sports car, but as a key component in an owner's broader energy management strategy.

Sustainability is likely a key focus that extends beyond the electric powertrain. Audi could make extensive use of recycled and bio-based materials in the interior, aligning with the car's eco-friendly ethos. These materials wouldn't just be chosen for their environmental credentials, but also for their performance and aesthetic appeal. The result could be an interior that's both luxurious and sustainable, setting new standards for the industry.

The infotainment system in the electric TT is likely to be more advanced and intuitive than ever before. Voice command systems, gesture controls, and haptic feedback could all play a role in creating an interface that's both futuristic and user-friendly. The system could also feature advanced telemetry capabilities, enabling drivers to analyze their performance data in depth, as professional racing teams do.

Cybersecurity will be a crucial consideration in the tech-laden electric TT. As cars become more connected, they also become potential targets for digital threats. Audi will likely implement robust security measures to protect both the car's systems and the owner's data, ensuring that the TT remains a safe and secure platform.

Finally, the electric TT could feature advanced vehicle-to-vehicle (V2V) and vehicle-to-infrastructure (V2I) communication systems. These technologies could improve safety by allowing the car to anticipate and react to road conditions and other vehicles. On the track, this could translate into features like ghost-car racing, where drivers compete against virtual opponents or their own previous best laps.

In conclusion, the technology integration in the electric Audi TT has the potential to redefine the sports car experience. By blending cutting-edge tech with the TT's traditional strengths of style and performance, Audi could create a vehicle that's not just a car, but a rolling showcase of automotive innovation. This fusion of heritage and future tech could ensure the TT remains at the forefront of the sports car segment, even as it enters the electric age.

Section 12.5: Market Positioning and Competition

The transition to electric powertrains is reshaping the automotive landscape, and the sports car segment is no exception. As the Audi TT evolves into an electric vehicle, its market positioning and competitive landscape will undergo significant changes.

The electric sports car market is still in its infancy but growing rapidly. Pioneering models like the Porsche Taycan and Tesla Roadster have set new benchmarks for performance and technology, creating a new standard for what electric sports cars can achieve. These vehicles have demonstrated that electric powertrains can not only match but often surpass the performance of their combustion-engine counterparts, particularly in acceleration and instant torque delivery.

As Audi positions the electric TT in this emerging market, it will face competition from both established sports car manufacturers and new entrants specializing in electric vehicles. Traditional rivals like BMW are rumored to be developing electric successors to their hybrid sports cars, such as a potential i8 follow-up. These competitors will likely target a similar market segment, combining performance, style, and cutting-edge technology.

Audi's challenge will be to differentiate the electric TT in this increasingly crowded field. The company may choose to position the TT as a premium offering, justifying a higher price point through a combination of advanced technology, exceptional performance, and the cachet of the TT brand. This strategy would allow Audi to leverage the TT's storied history and loyal fan base while appealing to early adopters of electric vehicle technology who are willing to pay a premium for the latest innovations.

The Audi TT: A Journey Through Design and Innovation

The pricing strategy for the electric TT will be crucial. Audi must balance the need to recoup the significant investment in electric vehicle technology with the desire to make the car accessible to a broad range of potential buyers. One possible approach is to offer multiple variants of the electric TT to cater to different market segments. For instance, Audi might introduce a 'standard' version of the electric TT, focused on balancing performance with affordability, alongside a high-performance 'RS' variant that pushes the boundaries of electric sports car capability at a higher price point.

Marketing the electric TT will require a nuanced approach. Audi will need to appeal to both existing TT enthusiasts and a new generation of buyers who may be more interested in electric technology than in traditional sports-car values. The company could leverage its participation in Formula E racing to highlight the performance credentials of its electric powertrains, drawing a direct line from the race track to the road-going TT.

Audi might also emphasize the electric TT's role as a technology showcase for the brand. By positioning the TT as the pinnacle of Audi's electric vehicle technology, the company can use it to generate excitement and interest that could translate to sales across its entire EV lineup.

The potential for special editions and limited-run models also plays a significant role in the electric TT's market strategy. Exclusive variants help sustain interest in the model over time and allow Audi to test new technologies or design concepts in limited production runs before potentially incorporating them into the standard model.

As the automotive industry continues its shift towards electrification, the market positioning of the electric Audi TT will be critical to its success. By carefully balancing performance, technology, and brand heritage, Audi can establish the electric TT as a benchmark in the emerging electric sports car segment, continuing the model's legacy of innovation and design leadership into the electric era.

Section 12.6: Challenges and Opportunities

The transition of the Audi TT to an electric powertrain presents a unique set of challenges and opportunities that Audi must navigate carefully to ensure the model's continued success and relevance in the automotive landscape.

One of the primary challenges for an electric TT is overcoming range anxiety in a sports-car context. Unlike daily commuters, sports cars are often driven for pleasure on longer routes or taken to racetracks, where range becomes a critical factor.

To address this, Audi could develop a network of high-speed chargers strategically placed along popular driving routes and at well-known racetracks. This infrastructure would not only alleviate range concerns but also improve overall performance. Still, it could also become a unique selling point for the electric TT, positioning it as a sports car truly designed for enthusiast use in the electric era.

Maintaining the TT's character in electric form is another significant challenge. The TT has always been known for its responsive handling, engine note, and overall driving experience. Audi engineers will need to carefully tune the electric powertrain to recreate the agile, spirited feel that TT drivers have come to love. This could involve developing a unique torque delivery profile for the electric motors, fine-tuning the suspension to account for the EV's different weight distribution, and creating an artificial sound that captures the TT's personality without mimicking a traditional combustion engine.

Balancing performance and efficiency is a challenge that all electric vehicles face, but it's particularly crucial for a sports car like the TT. Buyers will expect exhilarating performance but also a practical range for both daily use and weekend getaways. Audi could address this by using advanced aerodynamics, incorporating active elements that deploy at higher speeds to reduce drag. Lightweight materials, such as carbon fiber reinforced plastics and aluminum alloys, could be used extensively to offset the weight of the battery pack. Additionally, Audi might explore a two-speed transmission, similar to that in the Porsche Taycan, to optimize both acceleration and high-speed efficiency.

One of the most significant challenges will be appealing to traditional sports car enthusiasts who are skeptical of an electric TT. These purists often value the mechanical nature of sports cars, including manual transmissions and the character of internal combustion engines.

To win over this crucial demographic, Audi could offer track experiences that showcase the electric TT's capabilities, allowing skeptics to experience firsthand its instant torque and unique driving dynamics. Furthermore, Audi might consider retaining some traditional sports car elements, such as a driver-focused cockpit and tactile controls, to bridge the gap between the familiar and the future.

However, the shift to an electric powertrain also presents numerous opportunities for innovation. Audi has the chance to leverage electric technology to introduce new features that were impossible or impractical with a combustion engine. For instance, the electric TT could offer a 'valet mode' that limits power output when others are driving the car, providing owners with peace of mind. The instant torque of electric motors can be used to implement an advanced traction control system that reacts faster than traditional systems, enhancing both performance and safety.

The electric TT could also feature a highly customizable driving experience. Different driving modes could dramatically alter the car's character, from a comfortable grand tourer to a track-focused performance machine, all at the touch of a button. This level of versatility could broaden the TT's appeal, attracting buyers who might have previously considered it too focused or impractical.

Moreover, the electric powertrain opens up new design possibilities. Without the need for a large front grille for cooling, designers have more freedom to create an aerodynamic front end.

The lack of exhaust systems allows for a cleaner rear design, potentially enhancing the TT's already aggressive silhouette.

In conclusion, while the challenges of creating an electric TT are significant, they are outweighed by the opportunities for innovation and improvement. If Audi can successfully address these challenges while capitalizing on the unique benefits of electric powertrains, the electric TT has the potential to not only continue the model's legacy but to redefine what a modern sports car can be in the electric age. The result could be a car that honors its heritage while boldly pushing into the future, setting new benchmarks for performance, efficiency, and desirability.

Section 12.7: The TT's Legacy and Future Impact

The Audi TT has long been a symbol of innovative design and performance, and its transition into the electric era presents an opportunity to reinforce and expand upon this legacy. As we look to the future, it's clear that the electric TT has the potential to not only honor its rich heritage but also to shape the future of sports cars in profound ways.

One of the most crucial aspects of the electric TT's development will be honoring its design heritage. The TT has always been known for its avant-garde styling, and this DNA must carry forward into its electric incarnation. Audi's designers face the exciting challenge of reinterpreting classic TT design elements for the electric age.

We might see nods to the original TT's clean, minimalist lines and bold wheel arches, reimagined with a futuristic twist. The signature rounded roofline could be subtly altered to improve aerodynamics without losing its essential character. By thoughtfully evolving these design cues, Audi can create a car that is unmistakably a TT while also being thoroughly modern and electric-appropriate.

Beyond design, the electric TT has the potential to set new benchmarks for electric sports cars. As one of the pioneers in this emerging segment, Audi has the opportunity to define what an electric sports car can be. This could involve innovations in battery technology, pushing the boundaries of range and charging speed. The TT could showcase new approaches to weight distribution and chassis design that optimize the unique characteristics of electric powertrains. These innovations could influence the broader sports car market, setting new standards for performance, efficiency, and driving dynamics in the EV era.

The role of the TT in shaping Audi's brand image cannot be overstated. As Audi transitions to an all-electric lineup, the TT will be a key player in demonstrating that the brand's core values of progress through technology and exceptional design remain intact. A successful electric TT could bolster Audi's reputation as a leader in both design and technology, proving that the shift to electric power doesn't mean sacrificing the passion and performance that automotive enthusiasts crave. It could serve as a halo car for Audi's electric efforts, drawing attention and excitement to the brand's broader EV lineup.

The development of the electric TT is likely to yield technologies and innovations with far-reaching impacts across Audi's model range. For instance, high-performance electric motors developed for the TT could find their way into other Audi sports models. Optimizing battery technology for the unique demands of a sports car could lead to breakthroughs that improve the entire Audi EV lineup. Even software innovations, such as advanced torque vectoring systems or performance-oriented drive modes, could trickle down to other models, enhancing the driving experience across the board.

Finally, the transition to electric power offers the TT a unique place in automotive history. The original TT was pioneering in its design and helped establish Audi as a design leader. By successfully embracing electric technology, the TT has the opportunity to return to the forefront of automotive innovation. It could be remembered as a model that successfully bridged two eras of automotive technology, maintaining its core identity while adapting to the demands of a new age.

The electric TT, therefore, represents more than just the evolution of a single model. It symbolizes the broader transformation of the sports car for a new era of sustainable performance. As it moves into its electric future, the TT has the potential to capture the imagination of car enthusiasts worldwide once again, proving that the thrill of driving and the push for sustainability can go hand in hand. In doing so, it may secure its place not only in Audi's lineup but also in the pantheon of all-time great sports cars.

ABOUT THE AUTHOR

Todd Bandel is an accomplished author specializing in informational history books, particularly with a focus on the automotive industry. Drawing from 40 years of experience as an automotive technician, Todd combines deep expertise and passion to enlighten readers about the historical nuances of automobiles. Todd currently resides in San Diego, California, where he continues to explore and write about his enduring interest in automotive history.

Mechanicaddicts.com

www.ingramcontent.com/pod-product-compliance
Lightning Source LLC
Chambersburg PA
CBHW020656220526
45464CB00001B/457